# Lectures on Integral Transforms

Translations of

# MATHEMATICAL MONOGRAPHS

Volume 70

# Lectures on Integral Transforms

N. I. Akhiezer

American Mathematical Society
Providence, Rhode Island

# НАУМ ИЛЬИЧ АХИЕЗЕР

## ЛЕКЦИИ ОБ ИНТЕГРАЛЬНЫХ ПРЕОБРАЗОВАНИЯХ

ИЗДАТЕЛЬСТВО ПРИ
ХАРЬКОВСКОМ ГОСУДАРСТВЕННОМ УНИВЕРСИТЕТЕ
ИЗДАТЕЛЬСКОГО ОБЪЕДИНЕНИЯ «ВИЩА ШКОЛА»,
ХАРЬКОВ, 1984

Translated from the Russian by H. H. McFaden
Translation edited by Ben Silver

1980 *Mathematics Subject Classification* (1985 *Revision*).
Primary 44-02, 42B10; Secondary 44A35, 44A60, 42C05, 42C10, 42A82, 33A45, 33A65, 31B05.

**Abstract.** This monograph arose from lectures given by the author over the course of several years at Kharkov University for students in the Faculty of Mechanics and Mathematics. It is devoted to classical integral transforms, first and foremost the Fourier transform, and their applications. The general theory of the Fourier transform is constructed for the space $L^1(E_n)$ of integrable functions of $n$ variables, and the proof of the inversion theorem is based on the general Bochner theorem on integral transforms, a theorem that also has other applications within the framework of the book. The Fourier-Plancherel theory in the space $L^2(E_n)$ is studied. In addition to the general theory of integral transforms, connections are established with other areas of mathematical analysis such as the theory of harmonic and analytic functions, the theory of orthogonal polynomials, and the moment problem and other problems. Applications of integral transforms to problems in mathematical physics are considered. The book is intended for a broad circle of mathematicians and physicists interested in integral transforms.

---

**Library of Congress Cataloging-in-Publication Data**

Akhiezer, N. I. (Naum Il′ich), 1901–1980
[Lekt͡sii ob integral′nykh preobrazovanii͡akh, English]
Lectures on integral transforms/N. I. Akhiezer
vi, 107 pp. 23.5 cm. – (Translations of mathematical monographs; v. 70)
Translation of: Lekt͡sii ob integral′nykh preobrazovanii͡akh.
ISBN 0-8218-4524-1
1. Integral transforms. I. Title. II. Series.
QA432.A5313 1988 88-19393
515.7′23–dc19 CIP


The paper used in this book is acid-free and falls within the guidelines
established to ensure permanence and durability. ∞
This publication was typeset using $\mathcal{AMS}$-TEX,
the American Mathematical Society's TEX macro system.

10 9 8 7 6 5 4 3 2 96 95 94 93 92

# Contents

CHAPTER 1

# Averaging Operators and the Bochner Theorem

**1.** A linear integral operator is defined by the formula

$$g(x) = \int_G T(x,t)f(t)\,dt = (Tf)(x),$$

where $G$ is an open set in the $n$-dimensional Euclidean space $E_n$, $x = \{x_1, \dots, x_n\}$ and $t = \{t_1, \dots, t_n\}$ are points (or vectors), and $dt = dt_1 \cdots dt_n$ is the volume element of $E_n$. The function $T(x,t)$ is fixed, and is called the *kernel*(1) of the operator; the function $f$ runs through a certain collection $D_T$, the so-called *domain of the operator* (or *transform*) $T$, and $g$ also runs through some collection $\Delta_T$ called the *range of* $T$. The most important operator of this kind is the Fourier transform, which is discussed in courses in mathematical analysis. The Fourier transform, as well as certain other integral transforms, finds applications in the theory of linear equations in cases when the integral transform can be used to carry an equation in $D_T$ into an equation in $\Delta_T$ that is more convenient for solution or investigation.

**2.** Integral operators include the *smoothing* or *averaging operator*, which has various applications in mathematical analysis.

Let $f$ be an arbitrary measurable function on $G$. We can extend it by zero outside $G$ and thus assume that $f$ is defined everywhere in $E_n$. It will be assumed that $f$ is locally Lebesgue integrable. Then in considering the function in a neighborhood of a point $x$ it is natural to introduce the function's mean value in a ball about $x$ or on a sphere about $x$. If the radius is $\delta$, then the ball is denoted by $K_n^x(\delta)$ and the sphere by $S_{n-1}^x(\delta)$. The following notation is also used below: $K_n(\delta) = K_n^0(\delta)$, $S_{n-1}(\delta) = S_{n-1}^0(\delta)$, $K_n = K_n(1)$, and $S_{n-1} = S_{n-1}(1)$. The volume of $K_n^x(\delta)$ is equal to $|K_n(\delta)| = (\sqrt{\pi})^n\delta^n/\Gamma(n/2+1)$, and the area of the sphere

(1)The kernel is often denoted by the letter $K$ or $k$ (German "Kern"). If $T(x,t) = K(x-t)$, then the function $K(z)$ giving rise to this kernel is also called the kernel.

$S_{n-1}^x(\delta)$ is $|S_{n-1}\delta| = 2(\sqrt{\pi})^n\delta^{n-1}/\Gamma(n/2)$.[2] The results of these averagings are denoted by

$$f_\delta(x) = \frac{1}{|K_n(\delta)|}\int_{K_n^x(\delta)} f(y)\,dy, \quad \tilde{f}_\delta(x) = \frac{1}{|S_{n-1}(\delta)|}\int_{S_{n-1}^x(\delta)} f(y)\,d\sigma,$$

where $d\sigma$ is the area element of the sphere.

By the classical Lebesgue theorem, for almost all $x \in E_n$

$$\lim_{\delta\searrow 0}\frac{1}{|K_n(\delta)|}\int_{K_n^x(\delta)} f(y)\,dy = f(x). \tag{1}$$

This fact is proved in mathematical analysis courses for $n = 1$. The cube about $x$ with edge $\delta$ can be taken instead of the ball $K_n^x(\delta)$. If the cube is replaced by a rectangular parallelepiped, then the situation becomes more complicated.[3] The assertion (1) can be rewritten in the form

$$\int_{K_n^x(\delta)} [f(y) - f(x)]\,dy = o(\delta^n) \qquad (\delta \to 0). \tag{1$'$}$$

We now introduce the quantities

$$\tilde{\omega}_x(\delta) = \int_{S_{n-1}^x(\delta)} |f(y) - f(x)|\,d\sigma,$$

and

$$\omega_x(\delta) = \int_{K_n^x(\delta)} |f(y) - f(x)|\,dy = \int_0^\delta \tilde{\omega}_x(\rho)\,d\rho.$$

The following proposition, which strengthens (1′), is also due to Lebesgue.

---

(2) These two quantities are connected by the relation

$$\frac{d}{d\delta}|K_n(\delta)| = |S_{n-1}(\delta)|.$$

(3) We have the following theorem (Marcinkiewicz, Zygmund). Suppose that the rectangular parallelepiped $Q_n^x(h)$ about $x$ has edges of length $h_1, \ldots, h_n$; if for some $M \geq 1$ and for $h \to 0$

$$\frac{1}{M} \leq \frac{h_1}{h_k} \leq M \qquad (k = 1, 2, \ldots, r; r \leq n)$$

and if

$$\int |f(y)|\{\ln^+|f(y)|\}^{n-r}\,dy < \infty \tag{$*$}$$

locally, then

$$\lim_{h\to 0}\frac{1}{|Q_n(h)|}\int_{Q_n^x(h)} f(y)\,dy = f(x) \tag{$**$}$$

a.e. Examples show that if $(*)$ does not hold, then $(**)$ can fail to hold a.e.

THEOREM 1. *For almost all $x \in E_n$*

$$\omega_x(\delta) = o(\delta^n) \qquad (\delta \to 0). \tag{2}$$

DEFINITION. The points $x$ at which (2) holds are called *Lebesgue points* of $f$.

PROOF. Assuming for simplicity of the formulas that $f$ is real, and denoting by $\{r^{(m)}\}_1^\infty$ the set of all rational numbers, we write for any $m$

$$\begin{aligned}\omega_x(\delta) &\le \int_{K_n^x(\delta)} |f(y) - r^{(m)}|\,dy + \int_{K_n^x(\delta)} |f(x) - r^{(m)}|\,dy \\ &= \int_{K_n^x(\delta)} |f(y) - r^{(m)}|\,dy + |K_n(\delta)| \cdot |f(x) - r^{(m)}|.\end{aligned}$$

By relation (1) the first term on the right-hand side is $O(\delta^n) \cdot |f(x) - r^{r(m)}|$ as $\delta \to 0$ for almost all $x \in E_n$. Consequently, for almost all $x \in E_n$

$$\overline{\lim_{\delta \to 0}} \frac{1}{\delta^n} \omega_x(\delta) = O_x(1) \cdot |f(x) - r^{(m)}|. \tag{3}$$

For a given $m$ relation (3) is valid for all $x \in E_n \backslash e^{(m)}$, where $e^{(m)}$ is a set of measure zero in $E_n$. If we set $e = \bigcup_m e^{(m)}$, so that $e$ also has measure zero, then (3) holds in $E_n \backslash e$ for any $m$. However, for each $x \in E_n$ there is an $m$ such that $|f(x) - r^{(m)}| < \varepsilon$. Therefore,

$$0 \le \overline{\lim_{\delta \to 0}} \frac{1}{\delta^n} \omega_x(\delta) \le O_x(1) \cdot \varepsilon$$

for $x \in E_n \backslash e$. But $\varepsilon > 0$ is arbitrary; consequently,

$$\lim_{\delta \to 0} \frac{1}{\delta^n} \omega_x(\delta) = 0.$$

**3.** THEOREM 2. *Suppose that $K(x) \in L^1(E_n)$ and $\int_{E_n} K(x)\,dx = 1$. Then*

$$f(x) = \lim_{\varepsilon \searrow 0} \int_{E_n} b(y) \frac{1}{\varepsilon^n} K\left(\frac{y - x}{\varepsilon}\right) dy$$

*at each point $x$ where the bounded function $f$ is continuous.*

REMARK. This theorem was used as far back as Weierstrass under certain restrictive assumptions. The integral on the right-hand side can represent a function of $x$ that has a certain smoothness or is even analytic, depending on the kernel $K(x)$. Weierstrass used Theorem 2 to prove the famous approximation theorem that bears his name.

PROOF. Setting $y = x + \varepsilon z$, we find that

$$\int_{E_n} f(y) \frac{1}{\varepsilon^n} K\left(\frac{y - x}{\varepsilon}\right) dy = \int_{E_n} K(z) f(x + \varepsilon z)\,dz.$$

If $\sup|f| = M$, then $|K(z)\cdot f(x+\varepsilon z)| \leq M\cdot|K(z)|$ for the integrand, i.e., it has an integrable majorant. If $x$ is a point of continuity, then as $\varepsilon \to 0$ the integrand has the limit $K(z)\cdot f(z)$, and, by the Lebesgue dominated convergence theorem,

$$\lim_{\varepsilon\searrow 0}\int_{E^n} K(z)\cdot f(x+\varepsilon z)\,dz = f(x)\int_{E_n} K(z)\,dz = f(x).$$

**4.** A deficiency of Theorem 2 is the assumption that $f$ is bounded. If we want to consider increasing functions, then it is necessary to take kernels that decrease sufficiently rapidly.

DEFINITION. A *kernel of Fejér type* is defined to be a measurable function $k(z)$ such that $(1+|z|)^{n+\lambda}\cdot|k(z)| \leq M < \infty$ for some $\lambda > 0$, and $\int_{E_n} k(z)\,dz = 1$.

REMARK. Among the kernels $k(z)$ we single out those for which $k(z) = k(|z|)$. They are called *radial kernels.*

THEOREM 3 (BOCHNER). ([4]) *If $k$ is a kernel of Fejér type and $f$ is a measurable function with*

$$\frac{f(z)}{(1+|z|)^{n+\lambda}} \in L^1(E_n),$$

*then at each Lebesgue point of $f$, hence almost everywhere,*

$$f(x) = \lim_{\varepsilon\searrow 0}\int_{E_n} f(y)\frac{1}{\varepsilon^n}k\left(\frac{y-x}{\varepsilon}\right)dy.$$

PROOF. Introducing the notation

$$T_\varepsilon(x) = \int_{E_n} f(y)\frac{1}{\varepsilon^n}k\left(\frac{y-x}{\varepsilon}\right)dy,$$

we find without difficulty that

$$T_\varepsilon(x) - f(x) = \int_{E_n}[f(y)-f(x)]\frac{1}{\varepsilon^n}k\left(\frac{y-x}{\varepsilon}\right)dy.$$

From this we get

$$\begin{aligned}|T_\varepsilon(x) - f(x) \leq & \int_{|y-x|<\varepsilon}|f(y)-f(x)|\cdot\frac{1}{\varepsilon^n}k\left|\left(\frac{y-x}{\varepsilon}\right)\right|dy \\ & + \int_{|y-x|>\varepsilon}^{|y-x|<\delta}|f(y)-f(x)|\cdot\frac{1}{\varepsilon^n}\cdot\left|k\left(\frac{y-x}{\varepsilon}\right)\right|dy \\ & + \int_{|y-x|>\delta}|f(y)-f(x)|\frac{1}{\varepsilon^n}\cdot\left|k\left(\frac{y-x}{\varepsilon}\right)\right|dy \\ = & I_1 + I_2 + I_3\end{aligned}$$

([4])We call Theorem 3 the *Bochner representation theorem.*

for any finite $\delta > \varepsilon$. In the first integral we use the fact that $|k| \leq M$. Therefore,

$$I_1 \leq \frac{M}{\varepsilon^n} \int_{K_n^x(\varepsilon)} |f(y) - f(x)|\, dy = \frac{M}{\varepsilon^n} \omega_x(\varepsilon) = o(1) \qquad (\varepsilon \to 0).$$

For the third integral we find that

$$I_3 \leq \frac{M}{\varepsilon^n} \int_{|y-x|>\delta} \frac{|f(y)| + |f(x)|}{|y-x|^{n+\lambda}} \varepsilon^{n+\lambda}\, dy = M\varepsilon^\lambda A(x, \delta).$$

The second integral remains:

$$\begin{aligned}
I_2 &\leq \frac{M}{\varepsilon^n} \int_{|y-x|>\varepsilon}^{|y-x|<\delta} |f(y) - f(x)| \cdot \frac{\varepsilon^{n+\lambda}\, dy}{|y-x|^{n+\lambda}} \\
&= M\varepsilon^\lambda \int_\varepsilon^\delta \frac{d\rho}{\rho^{n+\lambda}} \int_{S_{n-1}^x(\rho)} |f(y) - f(x)|\, d\sigma \\
&= M\varepsilon^\lambda \int_\varepsilon^\delta \frac{d\rho}{\rho^{n+\lambda}} \tilde{\omega}_x(\rho) = M\varepsilon^\lambda \int_\varepsilon^\delta \frac{d}{d\rho}\omega_x(\rho) \cdot \frac{d\rho}{\rho^{n+\lambda}} \\
&= M\varepsilon^\lambda \left\{ \frac{\omega_x(\rho)}{\rho^{n+\lambda}} \bigg|_{\rho=\varepsilon}^{\rho=\delta} + (n+\lambda) \int_\varepsilon^\delta \frac{\omega_x(\rho)\, d\rho}{\rho^{n+\lambda+1}} \right\} \\
&\leq M\varepsilon^\lambda \left\{ \frac{1}{\varepsilon^\lambda} + (n+\lambda) \int_\varepsilon^\delta \frac{d\rho}{\rho^{\lambda+1}} \right\} \max_{0<\rho\leq\delta} \left[ \frac{\omega_x(\rho)}{\rho^n} \right] \\
&\leq M \left( 1 + \frac{n+\lambda}{\lambda} \right) \max_{0<\rho\leq\delta} \left[ \frac{\omega_x(\rho)}{\rho^n} \right] = o(1) \qquad (\delta \to 0).
\end{aligned}$$

Now we first find a $\delta > 0$ such that the term $I_2$ is sufficiently small, and then we let $\varepsilon$ go to 0.

**5.** As an example we consider the kernel due to Fejér himself. Here $n = 1$ and

$$k(z) = \frac{2}{\pi} \cdot \frac{\sin^2(z/2)}{z^2}.$$

Thus, $\lambda = 1$ in this case. We verify further that the kernel is normalized:

$$\frac{2}{\pi} \int_{-\infty}^{\infty} \frac{\sin^2(z/2)}{z^2}\, dz = \frac{1}{\pi} \int_{-\infty}^{\infty} (1 - \cos z) \left( -\frac{1}{z} \right)' dz = \frac{1}{\pi} \int_{-\infty}^{\infty} \frac{\sin z}{z}\, dz = 1.$$

For what is to follow we compute the integral

$$\frac{2}{\pi}\int_{-\infty}^{\infty}\frac{\sin^2(x/2)}{x^2}e^{-itx}\,dx = \frac{1}{\pi}\int_{-\infty}^{\infty}\frac{(1-\cos x)}{x^2}\cos tx\,dx$$
$$= \frac{1}{\pi}\left\{2\int_0^{\infty}\frac{\cos tx - 1}{x^2}\,dx - \int_0^{\infty}\frac{\cos(t-1)x-1}{x^2}\,dx - \int_0^{\infty}\frac{\cos(t+1)x-1}{x^2}\,dx\right\}.$$

In the second integral let $|t-1|x = u$, in the third $|t+1|x = u$, and in the first $|t|x = u$. We find that

$$\frac{2}{\pi}\int_{-\infty}^{\infty}\frac{\sin^2(x/2)}{x^2}e^{-itx}\,dx = \frac{2}{\pi}\int_0^{\infty}\frac{1-\cos u}{u^2}\,du\left[\frac{1}{2}|t+1| + \frac{1}{2}|t-1| - |t|\right]$$
$$= \begin{cases} 0 & (t \le -1), \\ 1-|t| & (-1 \le t \le 1), \\ 0 & (t \ge 1). \end{cases}$$

Suppose now that $f$ is a $2\pi$-periodic function integrable on an interval of length $2\pi$. We take $\varepsilon$ to be $1/m$, where $m$ is a positive integer, and let $T_\varepsilon(x) = \sigma_m(x)$. Thus,

$$\sigma_m(x) = \frac{2}{\pi}\int_{-\infty}^{\infty} f(y)\frac{\sin^2 m(y-x)/2}{m(y-x)^2}\,dy$$
$$= \frac{2}{\pi}\int_{-\infty}^{\infty} f\left(x+\frac{z}{m}\right)\cdot\sin^2\frac{z}{2}\cdot\frac{dx}{z^2}.$$

This function of $x$ clearly has period $2\pi$. The numbers

$$\frac{1}{2\pi}\int_0^{2\pi}\sigma_m(x)e^{-ikx}\,dx$$
$$= \frac{2}{\pi}\int_{-\infty}^{\infty}\frac{\sin^2(z/2)}{z^2}\,dz\cdot\frac{1}{2\pi}\int_0^{2\pi} f\left(x+\frac{z}{m}\right)e^{-ikx}\,dx$$
$$= \frac{2}{\pi}\int_{-\infty}^{\infty}\frac{\sin^2(z/2)}{z^2}\,dz\cdot\frac{1}{2\pi}\int_0^{2\pi} f(t)e^{-ikt}e^{ikz/m}\,dt$$
$$= c_k\cdot\frac{2}{\pi}\int_{-\infty}^{\infty}\frac{\sin^2(z/2)}{z^2}e^{izk/m}\,dz$$

are the Fourier coefficients of the function $\sigma_m(x)$, where the $c_k$ are the Fourier coefficients of $f$. On the basis of the above computations,

$$\frac{1}{2\pi}\int_0^{2\pi}\sigma_m(x)e^{-ikx}\,dx = \begin{cases} c_k(1-|k|/m) & (0 \le |k| \le m), \\ 0 & (|k| \ge m). \end{cases}$$

Thus, only the first $m$ harmonics of the function $\sigma_m(x)$ are nonzero. Consequently,

$$\sigma_m(x) = \sum_{k=-(m-1)}^{m-1} \left(1 - \frac{|k|}{m}\right) c_k e^{ikx},$$

i.e., $\sigma_m(x)$ is a trigonometric sum of order $m-1$.([5]) It is not hard to verify that

$$\sigma_m(x) = \frac{s_0 + s_1(x) + \cdots + s_{m-1}(x)}{m},$$

where

$$s_k(x) = \sum_{p=-k}^{k} c_p e^{ipx}$$

is the $k$th partial Fourier sum of $f$.

It follows from the general Bochner theorem applied to this case that

$$\lim_{m\to\infty} \sigma_m(x) = f(x) \quad \text{at each Lebesgue point of } f.$$

It is possible to get stronger results if we use the equality

$$\sigma_m(x) - f(x) = \frac{2}{\pi} \int_{-\infty}^{\infty} \left[f\left(x + \frac{z}{m}\right) - f(x)\right] \cdot \frac{\sin^2(z/2)}{z^2}\, dz.$$

1) Let $f$ be bounded ($|f| \leq M$) and continuous in the interval $(a, b)$. In this case the sequence $\sigma_m(x)$ tends uniformly to $f(x)$ in each closed interval $[a+\gamma, b-\gamma] \subset (a, b)$ as $m \to \infty$. For the proof, note that

$$|\sigma_m(x) - f(x) \leq \frac{2}{\pi} \int_{-A}^{A} \left|f\left(x + \frac{z}{m}\right) - f(x)\right| \cdot \frac{\sin^2(z/2)}{z^2}\, dz$$
$$+ \frac{2}{\pi} \int_{|z|>A} \left|f\left(x + \frac{z}{m}\right) - f(x)\right| \frac{\sin^2(z/2)}{z^2}\, dz.$$

The second term has the estimate

$$\frac{2}{\pi} \int_{|z|>A} \left|f\left(x + \frac{z}{m}\right) - f(x)\right| \frac{\sin^2(z/2)}{z^2}\, dz \leq \frac{8M}{\pi} \int_{A}^{\infty} \frac{dz}{z^2} = \frac{8M}{\pi A},$$

and hence the second term is less than $\varepsilon > 0$ for all $x$ if $A$ is sufficiently large. On the other hand, $|f(x + z/m) - f(x)| < \varepsilon$ if $x \in [a+\gamma, b-\gamma]$ and $|z| \leq A$ for fixed $A$ and for all sufficiently large $m$. Hence, $|\sigma_m(x) - f(x)| < 2\varepsilon$ $(a + \gamma \leq x \leq b - \gamma)$ for all sufficiently large $m$.

---

([5])It is called a *Fejér sum*.

2) If $f$ is continuous everywhere and satisfies a Cauchy-Lipschitz condition of order $\lambda$ $(0 < \lambda < 1)$ (so that $|f(x+h) - f(x)| < Mh^{\lambda}$ for any $x$ and any $h > 0$), then $|\sigma_m(x) - (x)| \leq C_{\lambda} M/m^{\lambda}$, with

$$C_{\lambda} \leq \frac{4}{\pi} \int_0^{\infty} \frac{\sin^2(z/2)}{z^{2-\lambda}}\, dz < \frac{3}{1-\lambda}.$$

The proof is left to the reader.

**6.** PROBLEMS. 1. If $f$ satisfies a Cauchy-Lipschitz condition of order 1, then

$$|\sigma_m(x) - f(x)| \leq CM \frac{\ln m}{m}.$$

2. Suppose that $\max_{0 \leq x \leq 2\pi} |\sigma_m(x) - f(x)| = o(1/m)$ for $m \to \infty$. Prove that $f(x) = \text{const}$ in this case.

3. Suppose that $f$ satisfies a Cauchy-Lipschitz condition of order 1 on some compact set $G \subset E_n$, i.e., $|f(y) - f(x)| \leq L \cdot |y - x|$ $\forall x, y \in G$. Prove that $f$ can be extended to the whole of $E_n$ in such a way that the resulting function is in the same class with the same constant $L$.

*Hint.* Assuming that $f$ is real, introduce in $E_n$ the function $\Phi_y(x) = f(y) + L \cdot |y - x|$ for any point $y \in G$, and then let $\Phi(x) = \min_{y \in G} \Phi_y(x)$.

4. Define the kernel $\tilde{k}(z)$ on $E_1$ by

$$\tilde{k}(z) = \begin{cases} \dfrac{1}{\pi} \cdot \dfrac{z}{z^2+1} & (-1 \leq z \leq 1), \\[2ex] \dfrac{1}{\pi}\left(\dfrac{z}{z^2+1} - \dfrac{1}{z}\right) & (|z| > 1). \end{cases}$$

Thus, the condition

$$(1 + |z|)^{1+\lambda} \cdot |\tilde{k}(z)| \leq M < \infty$$

holds even with $\lambda = 2$. However, the condition $\int_{-\infty}^{\infty} \tilde{k}(z)\, dz \neq 0$ does not hold. Using the same technique as in part 4, prove the following proposition:

*If $f(x)$ is a measurable function $(-\infty < x < \infty)$ and*

$$\int_{-\infty}^{\infty} \frac{|f(x)|}{1 + |x|}\, dx < \infty,$$

*then at each Lebesgue point $x$ of $f$*

$$\lim_{\varepsilon \searrow 0} \int_{-\infty}^{\infty} f(t) \cdot \frac{1}{\varepsilon} \cdot \tilde{k}\left(\frac{x-t}{\varepsilon}\right) dt = 0.$$

CHAPTER 2

# The Fourier Transform in $L^1$

**1.** In this chapter we begin a study of trigonometric integrals. These integrals will be written in the form

$$\frac{1}{\sqrt{2\pi}} \int_{-\infty}^{\infty} f(x) e^{ixt}\, dx$$

in the case of functions $f(x)$ of a single variable $x \in (-\infty, \infty)$, where $t$ is a real variable. Of course, we have to indicate here the sense in which the integral is understood. The situation is simplest when $f \in L^1(-\infty, \infty)$, i.e., $f$ is measurable and the integral of its modulus (in the Lebesgue sense) is finite. The formula written above defines on the whole space $L^1(-\infty, \infty)$ an additive and homogeneous operator called the *Fourier transform* in $L^1(-\infty, \infty)$. The factor $1/\sqrt{2\pi}$ is not essential, but sometimes turns out to be convenient, as we shall later see. We introduce at once the Fourier transform in $L^1(E_n)$ for arbitrary $n$. For this, denote the inner product in $E_n$ by $(x, t) = x_1 t_1 + \cdots + x_n t_n$. We also introduce the following extended notation for the Fourier transform:

$$\hat{f}(t) = \frac{1}{(\sqrt{2\pi})^n} \int_{E_n} f(x) e^{-i(x,t)}\, dx, \tag{1}$$

where $f \in L^1(E_n)$.

Some properties of the Fourier transform are almost obvious:

1) $\hat{f}(t)$ is bounded, namely, $|\hat{f}(t)| \leq \|f\|_1/(\sqrt{2\pi})^n$ for any $t \in E_n$, where $\|f\|_1 = \int_{E_n} |f(x)|\, dx$ is the norm of $f$ in $L^1(E_n)$.

1′) If a sequence $\{f_k\}_1^\infty$ converges in the $L^1(E_n)$-metric to $f$, then $\{\hat{f}_k\}_1^\infty$ converges uniformly to $\hat{f}$ (this follows from 1)).

2) $\hat{f}(t)$ is uniformly continuous in $E_n$.

3) $\hat{f}(t)$ tends to 0 as $|t| \to \infty$ (this statement is called the Riemann-Lebesgue lemma).

To prove assertion 2) we represent $\hat{f}(t)$ in the form

$$\hat{f}(t) = \frac{1}{(\sqrt{2\pi})^n} \int_{|x|<A} f(x) e^{-i(x,t)}\, dx + \frac{1}{(\sqrt{2\pi})^n} \int_{|x|>A} f(x) e^{-i(x,t)}\, dx.$$

The second integral can be estimated as follows:

$$\left|\frac{1}{(\sqrt{2\pi})^n}\int_{|x|>A} f(x)e^{-i(x,t)}\,dx\right| \leq \frac{1}{(\sqrt{2\pi})^n}\int_{|x|>A} |f(x)|\,dx.$$

The right-hand side is smaller than an arbitrarily small $\varepsilon > 0$ for sufficiently large $A$. It is obvious that the first integral is uniformly continuous with respect to $t \in E_n$.

To prove the Riemann-Lebesgue lemma we use the fact that the set of simple compactly supported functions is dense in $L^1(E_n)$. Recall that a *simple* function is defined to be a function taking only finitely many distinct values.([6]) The characteristic function $\kappa(x)$ of a rectangular parallelepiped with edges parallel to the coordinate axes is a simple function. The Riemann-Lebesgue lemma is clear for such a function, since in this case

$$\begin{aligned}\hat{\kappa}(t) &= \frac{1}{(\sqrt{2\pi})^n}\int_{\alpha_1}^{\beta_1}\cdots\int_{\alpha_n}^{\beta_n} e^{-i(t,x)}\,dx_1\cdots dx_n \\ &= \frac{i^n}{(\sqrt{2\pi})^n}\prod_{k=1}^{n}\frac{e^{-i\beta_k t_k}-e^{-i\alpha_k t_k}}{t_k},\end{aligned}$$

and hence $|\hat{\kappa}(t)| < C/|t|$ for $|t| \geq 1$, where $C$ does not depend on $t$, because $t_k^2 \geq |t|^2/n$ always holds for at least one $k$.

On the other hand, the linear span of such characteristic functions is already dense in $L^1(E_n)$. This implies the Riemann-Lebesgue lemma in full generality.

2. **Definition.** If both functions $f$ and $g$ belong to $L^1(E_n)$, then their *convolution* is defined to be the integral

$$\int_{E_n} f(y)g(x-y)\,dy = (f * g)(x). \tag{2}$$

We prove that the integral in (2) exists([7]) for almost all $x \in E_n$ and represents a function in $L^1(E_n)$. First note that the function $f(y)g(x-y)$ is measurable in $E_n \times E_n$. Next, we write the equality

$$\int_{E_n} |f(y)|\cdot|g(x-y)|\,dx = |f(y)|\cdot\|g\|_1,$$

---

([6]) *Editor's remark.* Functions taking at most countably many distinct values are sometimes regarded as simple functions.

([7]) The integral in (2) can fail to exist for some values of $x$. Here is a simple example for $n = 1$: $f(y) = g(y) = 1/\sqrt{y}(1+|y|)$. Then for $x = 0$ the integral is

$$\int_{-\infty}^{\infty}\frac{dy}{|y|\cdot(1+|y|)^2} = \infty.$$

whose right-hand side is in $L^1(E_n)$. Consequently, the iterated integral

$$\int_{E_n} dy \int_{E_n} |f(y)| \cdot |g(x-y)|\, dx$$

is finite. But in this case Fubini's theorem gives us that, first, the double integral

$$\iint_{E_n \times E_n} |f(y)||g(x-y)|\, dx\, dy$$

is finite, second, the integral

$$\int_{E_n} |f(y)| \cdot |g(x-y)|\, dy \tag{2'}$$

is finite for almost all $x$ and belongs to $L^1(E_n)$, and, third, all three integrals (the double one and the two iterated ones) have the same value. Moreover, all three integrals have the same finite value with $f$ and $g$ instead of $|f|$ and $|g|$. Our assertion follows from what has been noted about the integral (2′).

We remark further that $\|f * g\|_1 \le \|f\|_1 \cdot \|g\|_1$. It is also clear that $f * g = g * f$.

Let $f, g \in L^1(E_n)$, and let $h = f * g$. In this case

$$\hat{h}(t) = (\sqrt{2\pi})^n \hat{f}(t) g(t). \tag{3}$$

Indeed,

$$\hat{h}(t) = \frac{1}{(\sqrt{2\pi})^n} \int_{E_n} e^{-i(t,x)}\, dx \int_{E_n} f(y) g(x-y)\, dy.$$

By Fubini's theorem,

$$h(t) = \frac{1}{(\sqrt{2\pi})^n} \int_{E_n} f(y)\, dy \int_{E_n} g(x-y) e^{-i(t,x)}\, dx.$$

Performing the change $x = y + z$, we find that

$$\hat{h}(t) = \frac{1}{(\sqrt{2\pi})^n} \int_{E_n} f(y) e^{-i(t,y)}\, dy \int_{E_n} g(z) e^{-i(t,z)}\, dz,$$

and this is the required relation (3).

Similarly, applying Fubini's theorem to the integral

$$\frac{1}{(\sqrt{2\pi})^n} \int_{E_n} f(x)\, dx \int_{E_n} g(y) e^{-i(x,y)}\, dy,$$

we find that for any $f, g \in L^1(E_n)$

$$\int_{E_n} f(x) \hat{g}(x)\, dx = \int_{E_n} g(y) \hat{f}(y)\, dy. \tag{4}$$

The following properties of the Fourier transform are easy to verify:

1) If $F(x) = f(Rx)$, where $R$ is a real number, then

$$\hat{F}(t) = \frac{1}{|R|^n}\hat{f}\left(\frac{t}{R}\right).$$

2) If $F(x) = f(x + c)$, where $c$ is a constant vector, then

$$\hat{F}(t) = e^{i(t,c)} \cdot \hat{f}(t).$$

3) If $F(x) = f(x)e^{-i(x,c)}$, then $\hat{F}(t) = \hat{f}(t + c)$.

4) If $f(x) = f_1(x_1) \cdots f_n(x_n)$, where $f_k(x_k) \in L^1(E_1)$ $(k = 1, \dots, n)$, then $\hat{f}(t) = \hat{f}_1(t_1) \cdots \hat{f}_n(t_n)$.

We also prove the proposition

5) If $f$ is a radial function, i.e., $f(x) \equiv f(|x|)$, then so is $\hat{f}$.

A function is *radial* if and only if it does not change when $E_n$ undergoes a rotation about 0, i.e., it is invariant under any unitary operator $U$ on $E_n$: $f(x) = f(Ux)$. Assuming that $f$ has this property, we can write the following equality:

$$\hat{f}(t) = \frac{1}{(\sqrt{2\pi})^n}\int_{E_n} f(x)e^{-i(t,x)}\,dx = \frac{1}{(\sqrt{2\pi})^n}\int_{E_n} f(Ux)e^{-i(t,x)}\,dx.$$

If $x' = Ux$, then $dx' = dx$ and $(t, x) = (Ut, x')$. Therefore,

$$\hat{f}(t) = \frac{1}{(\sqrt{2\pi})^n}\int_{E_n} f(x')e^{-i(Ut,x')}\,dx' \equiv \hat{f}(Ut),$$

and hence $\hat{f}(t) = \hat{f}(|t|)$.

**3.** In this section we find the Fourier transform of the function $e^{-|x|^2/2}$ $(x \in E_n)$ as an important example. Since $e^{-|x|^2/x} = \prod_1^n e^{-x_k^2/2}$, assertion 4) of §2 applies here, and we can confine ourselves to finding the Fourier transform in $E_1$ of the function $g(x) = e^{-x^2/2}$ $(-\infty < x < \infty)$. Starting from the formula

$$\hat{g}(t) = \frac{1}{\sqrt{2\pi}}\int_{-\infty}^{\infty} e^{-x^2/2}e^{-itx}\,dx = \sqrt{\frac{2}{\pi}}\int_0^{\infty} e^{-x^2/2}\cos tx \cdot dx$$

we differentiate with respect to $t$ and then integrate by parts. We get

$$\begin{aligned}\frac{d}{dt}\hat{g}(t) &= -\sqrt{\frac{2}{\pi}}\int_0^{\infty} xe^{-x^2/2}\sin tx \cdot dx \\ &= \sqrt{\frac{2}{\pi}}\left\{e^{-x^2/2}\sin tx\,\Big|_{x=0}^{x=\infty} - t^2\int_0^{\infty} e^{-x^2/2}\cos tx \cdot dx\right\} = -t\hat{g}(t).\end{aligned}$$

Hence, $\hat{g}(t) = Ce^{-t^2/2} = Cg(t)$. Since $g(0) = 1$, it follows that

$$C = \hat{g}(0) = \frac{1}{\sqrt{2\pi}} \int_{-\infty}^{\infty} e^{-x^2/2}\, dx = \sqrt{\frac{2}{\pi}} \int_0^{\infty} e^{-x^2/2}\, dx$$
$$= \frac{1}{\sqrt{\pi}} \int_0^{\infty} e^{-y} y^{-1/2}\, dy = \frac{1}{\sqrt{\pi}} \Gamma\left(\frac{1}{2}\right) = 1.$$

Thus, we have proved the equality

$$\frac{1}{\sqrt{(2\pi)}^n} \int_{E_n} e^{-|x|^2/2} e^{-i(t,x)}\, dx = e^{-|t|^2/2}.$$

This means that $e^{-|x|^2/2}$ is an eigenfunction of the Fourier transform in $L^1(E_n)$, and the corresponding eigenvalue is 1. Note that $e^{-|z|^2/2}/(\sqrt{2\pi})^n$ $(z \in E_n)$ satisfies all the conditions in the definition of a kernel of Fejér type, including normalization. For what follows, we call the function

$$W(z, \tau) = \frac{1}{(\sqrt{2\pi\tau})^n} e^{-|z|^2/2\tau} \tag{$5_1$}$$

a *Weierstrass* (or *Gauss-Weierstrass*) *kernel.* Here $\tau > 0$ is a parameter (introduced instead of $\varepsilon = \tau^{1/2}$). Further, we write the equality

$$W(z, \tau) = \frac{1}{(2\pi)^n} \int_{E_n} e^{-\tau|u|^2/2} e^{-i(u,z)}\, du, \tag{$5_2$}$$

which we shall have to use below.

**4.** As another important example we find the Fourier transform of the function $e^{-|x|}$ $(x \in E_n)$. Here we employ the formula([8])

$$\frac{1}{\sqrt{\pi}} \int_0^{\infty} e^{-u-c^2/4u} \frac{du}{\sqrt{u}} = e^{-c} \qquad (c \geq 0).$$

Using it with $c = |x|$, we find that

$$\frac{1}{(\sqrt{2\pi})^n} \int_{E_n} e^{-|x|} e^{-i(t,x)}\, dx$$
$$= \frac{1}{\sqrt{\pi}} \int_0^{\infty} e^{-u} \frac{du}{\sqrt{u}} \frac{1}{(\sqrt{2\pi})^n} \int_{E_n} e^{-|x|^2/4u} e^{-i(t,x)}\, dx.$$

---

([8])To prove this, let

$$g(c) = \frac{1}{\sqrt{\pi}} \int_0^{\infty} e^{-u-c^2/4u} \frac{du}{\sqrt{u}},$$

so that

$$g''(c) = -\frac{1}{\sqrt{\pi}} \int_0^{\infty} e^{-u-c^2/4u} \frac{c}{2u\sqrt{u}}\, du.$$

Setting $u = c^2/(4v)$ here, we get that

$$g''(c) = -\frac{1}{\sqrt{\pi}} \int_0^{\infty} e^{-c^2/4v-v} \frac{dv}{\sqrt{v}} = -g(c).$$

Consequently, $g(c) = Ae^{-c}$, and it remains to observe that $A = g(0) = 1$.

We make the change of variable $x = \sqrt{2u} \cdot y$ on the right-hand side and then use ($5_1$) and ($5_2$), getting that

$$\begin{aligned}
&\frac{1}{(\sqrt{2\pi})^n} \int_{E_n} e^{-|x|} e^{-i(t,x)}\, dx \\
&\quad = \frac{1}{\sqrt{\pi}} \int_0^\infty e^{-u} \frac{du}{\sqrt{u}} \left( \frac{\sqrt{2u}}{\sqrt{2\pi}} \right)^n \int_{E_n} e^{-|y|^2/2} e^{-i(t\sqrt{2u},y)}\, dy \\
&\quad = \frac{(\sqrt{2})^n}{\sqrt{\pi}} \int_0^\infty e^{-u-u|t|^2} u^{(n-1)/2}\, du \\
&\quad = \frac{2^{n/2}}{\sqrt{\pi}} \frac{1}{(1+|t|^2)^{(n+1)/2}} \int_0^\infty e^{-v} v^{(n-1)/2}\, dv \\
&\quad = \frac{2^{n/2}\Gamma((n+1)/2)}{\sqrt{\pi}} \cdot \frac{1}{(1+|t|^2)^{(n+1)/2}}.
\end{aligned}$$

The function

$$\frac{2}{|S_n|} \cdot \frac{1}{(1+|z|^2)^{(n+1)/2}} = \frac{1}{(2\pi)^n} \int_{E_n} e^{-|t|} e^{-i(t,z)}\, dt \tag{6}$$

also satisfies all the conditions in the definition of a Fejér kernel, including normalization, since([9])

$$\frac{2}{|S_n|} \int_{E_n} \frac{dz}{(1+|z|^2)^{(n+1)/2}} = 1.$$

The function

$$P(z,\eta) = \frac{2}{|S_n|} \cdot \frac{\eta}{(\eta^2+|z|^2)^{(n+1)/2}} \tag{$7_1$}$$

is called the *Poisson kernel*. Here the parameter $\eta > 0$ is written instead of $\varepsilon$. Formula (6) can be rewritten in the form

$$P(z,\eta) = \frac{1}{(2\pi)^n} \int_{E_n} e^{-\eta|t|} e^{-i(t,z)}\, dt. \tag{$7_2$}$$

---

([9])This integral is computed as follows. We have

$$\begin{aligned}
\frac{2}{|S_n|} \int_{E_n} \frac{dz}{(1+|z|^2)^{(n+1)/2}} &= \frac{2|S_{n-1}|}{|S_n|} \int_0^\infty \frac{r^{n-1}\, dr}{(1+r^2)^{(n+1)/2}} \\
&= \frac{\Gamma((n+1)/2)}{\Gamma(1/2)\Gamma(n/2)} \int_0^\infty \frac{t^{n/2-1}\, dt}{(1+t)^{(n+1)/2}} = 1.
\end{aligned}$$

However, by the inversion formula to be proved in the next chapter, (6) implies that

$$e^{-|t|} = \frac{2}{|S_n|} \int_{E_n} \frac{e^{i(t,z)}\, dz}{(1+|z|^2)^{(n+1)/2}},$$

from which the normalization is immediately clear.

**5. Problems.** 1. Let

$$\psi_{a,b}(t) = \sqrt{\frac{2}{\pi}} \cdot \frac{\cos at - \cos(a+b)t}{bt^2} \qquad (a > 0,\ b > 0).$$

Prove that $\hat{\psi}_{a,b}(x) = 1$ ($|x| \leq a$), $\hat{\psi}_{a,b}(x) = 0$ ($|x| \geq a+b$), and $\hat{\psi}_{a,b}(x)$ is linear on the remaining intervals.

2. Let $f(x) \in L^1(E_n)$ and $F(x) = x_k f(x) \in L^1(E_n)$, where $k$ is one of the numbers $1, \ldots, n$. Prove that $\hat{f}(t)$ has a continuous derivative with respect to $t_k$, and also that $\partial \hat{f}(t)/\partial t_k = -i\hat{F}(t)$.

3. Let $x = \{x_1, x'\}$, where $x' = \{x_2, \ldots, x_n\} \in E_{n-1}$. Suppose that $f(x)$ is absolutely continuous with respect to $x_1$ for almost all $x' \in E_{n-1}$. If $f(x) \in L^1(E_n)$ and $F(x) = \partial f(x)/\partial x_1 \in L^1(E_n)$, then $\hat{F}(t) = it_1 \hat{f}(t)$. Prove this assertion (in which $x_1$ and $t_1$ can be replaced by $x_k$ and $t_k$, of course).

4. Find the area of the unit sphere in $E_n$ by using the fact that the function $e^{-|x|^2/2}$ is invariant under the Fourier transform.

5. Let

$$l(x) = \frac{1}{x} \int_0^\infty \frac{|\sin tx \cdot \sin t|}{t^2}\, dt \qquad (0 < x \leq 1).$$

Prove that

$$l(x) > \frac{2}{\pi} \ln \frac{1}{x} + \frac{3}{2} - \frac{3}{\pi}; \qquad l(x) < \frac{2}{\pi} \ln \frac{1}{x} + \frac{3}{2} + \frac{3}{\pi}.$$

*Hint.* Use the expansion

$$|\sin t| = \frac{4}{\pi} \sum_{k=1}^{\infty} \frac{1 - \cos 2kt}{4k^2 - 1}.$$

CHAPTER 3

# The Inversion Theorem in $L^1$. The Poisson Integral

**1.** In connection with each integral transformation the very first question arising concerns its inversion, i.e., the question of whether a function (inverse image) can be recovered from its transform, and, if that is possible, how. We consider this question in connection with the Fourier transform in $L^1(E_n)$.

THEOREM 1. *If* $f \in L^1(E_n)$ *and*

$$\hat{f}(t) = \frac{1}{(\sqrt{2\pi})^n} \int_{E_n} f(x) e^{-i(t,x)}\, dx, \tag{1}$$

*then at each Lebesgue point of* $f$

$$f(x) = \lim_{\tau \searrow 0} \frac{1}{(\sqrt{2\pi})^n} \int_{E_n} \hat{f}(t) e^{i(t,x)} e^{-\tau|t|^2/2}\, dt. \tag{2}$$

PROOF. We take the Weierstrass kernel in the Bochner representation theorem. Consider the quantity $T_\varepsilon(x)$ introduced in the proof of that theorem, let $\tau = \varepsilon^2$, and write $T_\tau(x)$ for simplicity instead of $T_\varepsilon(x) = T_{\sqrt{\tau}}(x)$. By $(5_1)$ and $(5_2)$ in Chapter 2,

$$\begin{aligned} T_\tau(x) &= \int_{E_n} f(y) \frac{1}{(\sqrt{2\pi\tau})^n} e^{-|y-x|^2/2\tau}\, dy \\ &= \frac{1}{(2\pi)^n} \int_{E_n} f(y)\, dy \int_{E_n} e^{-\tau|t|^2/2} e^{-i(t,y-x)}\, dt. \end{aligned}$$

In view of Fubini's theorem,

$$T_\tau(x) = \frac{1}{(2\pi)^n} \int_{E_n} e^{-\tau|t|^2/2} e^{i(t,x)}\, dt \int_{E_n} f(y) e^{-i(t,y)}\, dy,$$

which implies that

$$T_\tau(x) = \frac{1}{(\sqrt{2\pi})^n} \int_{E_n} \hat{f}(t) e^{i(t,x)} e^{-\tau|t|^2/2}\, dt. \tag{3}$$

Bochner's theorem gives us that $T_\tau(x) \to f(x)$ $(\tau \to 0)$ at all Lebesgue points of $f$, which is what was required.

REMARK. Comparing the $T_\tau(0)$ in the original expression for $T_\tau(x)$ with that following from (3), we get the equality

$$\int_{E_n} f(y)e^{-|y|^2/2\tau}\frac{dy}{\tau^{n/2}} = \int_{E_n} \hat{f}(t)e^{-\tau|t|^2/2}\,dt, \tag{4}$$

which is needed below.

Theorem 1 has an important corollary, usually called *the uniqueness theorem for the Fourier transform in $L^1(E_n)$*.

COROLLARY 1. *If two functions in $L^1(E_n)$ have the same Fourier transform, then they coincide almost everywhere.*

Because of the uniqueness theorem, the question of recovering a function $f$ from its Fourier transform makes sense. Formula $(2_1)$ gives one method of recovery. Another method is given (at each Lebesgue point of $f$) by the formula

$$f(x) = \lim_{\eta\searrow 0}\frac{1}{(\sqrt{2\pi})^n}\int_{E_n} \hat{f}(t)e^{i(x,t)}e^{-\eta|t|}\,dt, \tag{$2_2$}$$

the derivation of which is left to the reader (use the Poisson kernel).

COROLLARY 2. *If not only $f$ but also $\hat{f}$ is in $L^1(E_n)$, then the inversion formula*([10])

$$f(x) = \frac{1}{(\sqrt{2\pi})^n}\int_{E_n} \hat{f}(t)e^{i(x,t)}\,dt \tag{2}$$

*is valid.*

Indeed, if $\hat{f} \in L^1(E_n)$, then the integrand on the right-hand side of $(2_1)$ and $(2_2)$ has the absolutely integrable majorant $|\hat{f}(t)|$. Therefore, the Lebesgue dominated convergence theorem applies.

**2.** If the function $\hat{f}$ is not in $L^1(E_n)$, then the right-hand side of (2) can fail to make sense; in other words, the integral (2) can diverge. It is common to say that we "sum the divergent integral" with the help of formulas $(2_1)$ and $(2_2)$, assigning it a definite value. Here the general situation is as follows. The integral $\int_{E_n} F(x)\,dx$ diverges. Suppose that there exists a function (convergence factor) $\varphi(x,\varepsilon)$ with parameter $\varepsilon > 0$ such

([10])The reciprocal formulas (2) and (1) differ from each other only in that the coefficient of $i$ in the exponent in (2) has sign different from that in (1). Here we see the expediency of the factor $1/(\sqrt{2\pi})^n$ in front of the integral sign in the definition of the Fourier transform.

that $\varphi(x,\varepsilon) \to 1$ ($\varepsilon \searrow 0$), and suppose that the integral $\int_{E_n} F(x)\varphi(x,\varepsilon)\,dx$ converges for each $\varepsilon > 0$. If, furthermore, the limit

$$\lim_{\varepsilon\searrow 0} \int_{E_n} F(x)\varphi(x,\varepsilon)\,dx = J$$

exists and is finite, then we say that the integral $\int_{E_n} F(x)\,dx$ *converges in the generalized sense* (or *is summable*) to the value $J$. Hardy called this *convergence in the Pickwick sense*,(11) and wrote

$$(\mathrm{P}) \int_{E_n} F(x)\,dx = J.$$

Of course certain conditions must be satisfied for the use of this concept and of the "computation" of integrals based on it. For example, if the integral converges in the usual sense and also in the Pickwick sense, then its values must be the same in both cases. We can say that the integral (2) is summable to $f(x)$ at each Lebesgue point of $f$ both by the W-method (the Weierstrass method, formula $(2_1)$) and by the A-method (the Abel method, formula $(2_2)$). It is perhaps worth clarifying this concept by a simple example. The integral $\int_0^\infty e^{ix}\,dx$ is obviously divergent, since $\int_0^T e^{ix}\,dx = i(1-e^{iT})$ does not have a limit as $T \to \infty$. However,

$$\int_0^\infty e^{ix}e^{-\varepsilon x}\,dx = \frac{\varepsilon + i}{\varepsilon^2 + 1}$$

for $\varepsilon > 0$. Consequently, $\lim_{\varepsilon\searrow 0}\int_0^\infty e^{ix}e^{-\varepsilon x}\,dx = i$, which implies that $(\mathrm{A})\int_0^\infty \cos x\,dx = 0$, $(\mathrm{A})\int_0^\infty \sin x\,dx = 1$.

**3.** The following sufficient condition for $\hat f$ to belong to $L^1(E_n)$ is sometimes useful.

THEOREM 2. *Suppose that $f \in L^1(E_n)$ and the following conditions hold:*

1) *There exists a $\delta > 0$ such that $|f(x)| \le N < \infty$ for $|x| < \delta$ (i.e., $f(x)$ is bounded in some neighborhood of $x = 0$).*

2) *$\hat f(t) \ge 0$ for all $t \in E_n$.*

*Then $\hat f \in L^1(E_n)$.*

The proof is based on (4). We write it in the form

$$\int_{E_n} \hat f(t)e^{-\tau|t|^2/2}\,dt = \int_{|y|<\delta} f(y)e^{-|y|^2/2\tau}\frac{dy}{\tau^{n/2}} + \int_{|y|>\delta} f(y)e^{-|y|^2/2\tau}\frac{dy}{\tau^{n/2}}$$
$$= I_1 + I_2.$$

(11)After the character in Dickens' famous novel *The Pickwick papers.*

Then

$$|I_1| \le N \cdot \int_{|y|<\delta} e^{-|y|^2/2\tau} \frac{dy}{\tau^{n/2}} < N \int_{E_n} e^{-|y|^2/2\tau} \frac{dy}{\tau^{n/2}}$$
$$= N \int_{E_n} e^{-|u|^2/2}\, du = (\sqrt{2\pi})^n N,$$

and, on the other hand,

$$|I_2| \le e^{-\delta^2/2\tau} \cdot \frac{1}{\tau^{n/2}} \int_{|y|>\delta} |f(y)|\, dy \le e^{-\delta^2/2\tau} \cdot \frac{1}{\tau^{n/2}} \|f\|_1,$$

where the right-hand side is bounded for all $\tau > 0$. Thus,

$$\int_{E_n} \hat{f}(t) e^{-\tau|t|^2/2}\, dt \le M < \infty$$

for all $\tau > 0$. Since $\hat{f}(t) > 0$, this gives us that $\int_{E_n} \hat{f}(t)\, dt < \infty$.

**4.** The Weierstrass and Poisson kernels are connected with certain boundary value problems in mathematical physics. We consider the Poisson kernel, which plays an important role in the theory of harmonic functions.([12])

Recall that a *harmonic function* in some domain of a Euclidean space of two, three, or more dimensions is defined to be a twice continuously differentiable function $u$ on this domain such that $\Delta u = 0$ ($\Delta$ is the Laplace operator). As a domain we take the "upper half" of the space $E_{n+1}$, denoted by $E^+_{n+1}$ and defined by

$$E^+_{n+1} = \{(x, \eta) | x \in E_n,\ \ \eta > 0\} = E_n \times (0, \infty).$$

The harmonicity of a function $u(x, \eta)$ in $E^+_{n+1}$ means that for any $x \in E_n$ and any $\eta > 0$

$$\frac{\partial^2 u}{\partial x_1^2} + \frac{\partial^2 u}{\partial x_2^2} + \cdots + \frac{\partial^2 u}{\partial x_n^2} + \frac{\partial^2 u}{\partial \eta^2} = 0.$$

Taking some function $f \in L^1(E_n)$ and the Poisson kernel

$$P(z, \eta) = \frac{2}{|S_n|} \cdot \frac{\eta}{(|z|^2 + \eta^2)^{(n+1)/2}},$$

we can write the equality

$$\int_{E_n} f(y) P(x - y, \eta)\, dy = \frac{1}{(\sqrt{2\pi})^n} \int_{E_n} \hat{f}(t) e^{i(t,x)} e^{-\eta|t|}\, dt. \tag{5}$$

([12])The Weierstrass kernel plays an analogous role in the theory of heat conduction (see, for example, Problem 3 in Chapter 14).

Here we use formulas (4) and ($7_2$) of Chapter 2. The left-hand side of (5) is a function

$$u(x,\eta) = \frac{1}{|S_n|}\int_{E_n} f(y)\frac{\eta}{(\eta^2+|x-y|^2)^{(n+1)/2}}\,dy, \tag{6}$$

called the *Poisson integral for the half-space*.(13) The function $u(x,\eta)$ has derivatives of any order with respect to the variables $x_k$ and $\eta$, and they can be found by differentiating the kernel in (6) or (preferably) by differentiating the integrand on the right-hand side of (5). In this way we find that

$$\frac{\partial^2 u}{\partial x_k^2} = -\frac{1}{(\sqrt{2\pi})^n}\int_{E_n}\hat{f}(t)t_k^2 e^{i(t,x)}e^{-\eta|t|}\,dt \qquad (k=1,2,\dots,n)$$

and

$$\frac{\partial^2 u}{\partial \eta^2} = \frac{1}{(\sqrt{2\pi})^n}\int_{E_n}\hat{f}(t)|t|^2 e^{i(t,x)}e^{-\eta|t|}\,dt.$$

Addition of these expressions gives us that $u(x,\eta)$ is harmonic in $E_{n+1}^+$.

THEOREM 3. *If $f\in L^1(E_n)$, then the Poisson integral (6) is a harmonic function on $E_{n+1}^+$ with the following properties*:

a) $\lim_{\eta\searrow 0} u(x,\eta) = f(x)$ *at all Lebesgue points of* $f$.

b) $\int_{E_n}|u(x,\eta)|\,dx \le \|f\|_1$ $(\eta>0)$.

c) $\lim_{\eta\searrow 0}\int_{E^n}|u(x,\eta)-f(x)|\,dx = 0$.

PROOF. In connection with this case, property a) is the content of the Bochner representation theorem. Property b) can be proved with the help of Fubini's theorem:

$$\begin{aligned}\int_{E_n}|u(x,\eta)|\,dx &\le \frac{2}{|S_n|}\cdot\int_{E_n}dx\int_{E_n}|f(y)|\cdot\frac{\eta}{(\eta^2+|x-y|^2)^{(n+1)/2}}\,dy\\ &= \frac{2}{|S_n|}\cdot\int_{E_n}|f(y)|\,dy\int_{E_n}\frac{dz}{(1+|z|^2)^{(n+1)/2}} = \int_{E_n}|f(y)|\,dy.\end{aligned}$$

To prove c) we write

$$u(x,\eta)-f(x) = \frac{2}{|S_n|}\cdot\int_{E_n}[f(x+\eta z)-f(x)]\frac{dz}{(1+|z|^2)^{(n+1)/2}}.$$

(13)We later encounter the Poisson integral for a ball.

Again by Fubini's theorem,

$$\begin{aligned}\int_{E_n} &|u(x,\eta) - f(x)|\,dx \\ &\le \frac{2}{|S_n|}\cdot\int_{E_n}\frac{dz}{(1+|z|^2)^{(n+1)/2}}\int_{E_n}|f(x+\eta z) - f(x)|\,dx \\ &\le \frac{4}{|S_n|}\int_{|z|>A}\frac{dz}{(1+|z|^2)^{(n+1)/2}}\cdot\|f\|_1 \\ &\quad + \frac{2}{|S_n|}\cdot\int_{|z|<A}\frac{dz}{(1+|z|^2)^{(n+1)/2}}\int_{E_n}|f(x+\eta z) - f(x)|\,dx.\end{aligned}$$

The first term is as small as desired for sufficiently large $A$. On the other hand, since each $L^1(E_n)$-function is continuous in the $L^1(E_n)$-metric, the quantity $\int_{E_n}|f(x+\eta z) - f(x)|\,dx$ ($|z| < A$) converges uniformly to 0 as $\eta \to 0$.

An important role in the theory of harmonic functions is played by the so-called *Dirichlet problem* or *first boundary value problem*. It is formulated as follows in the classical theory: a bounded domain $G$ is given in space, and a continuous function $f$ is given on its boundary; we are required to find a function continuous in $\bar{G}$, harmonic in $G$, and coinciding with $f$ on the boundary of $G$. It is proved in the theory of partial differential equations that this problem has at most one solution, and the Dirichlet problem is solvable under certain general assumptions about the boundary. In essence, Theorem 3 concerns a generalization of the Dirichlet problem. Here the domain is unbounded, and the function given on its boundary is discontinuous in general. However, it is proved that there is a function harmonic in the domain and taking the given values in a generalized sense on the boundary ($\eta = 0$): almost everywhere according to property a), and in the $L^1(E_n)$-metric in view of property c). It should be mentioned that the theorem says nothing about the uniqueness of the solution of this Dirichlet problem. If we are concerned only with the boundary condition a), then uniqueness clearly fails to hold, because of any constant $C$ the harmonic function $u(x,\eta)+C\eta$ satisfies a) when $u(x,\eta)$ does. The following result should be considered in this connection.

THEOREM 4. *Suppose that $u(x,\eta)$ is harmonic in $E^+_{n+1}$ and satisfies the following conditions*:

b) $\int_{E_n}|u(x,\eta)|\,dx \le M < \infty$ *for each* $\eta > 0$.

c*) $\lim_{\eta'\searrow 0,\eta''\searrow 0}\int_{E_n}|u(x,\eta') - u(x,\eta'')|\,dx = 0$.

*Then $u(x, \eta)$ is representable by a Poisson integral, i.e., there exists a function $f \in L^1(E_n)$ such that*

$$u(x, \eta) = \frac{2}{|S_n|} \int_{E_n} f(y) \frac{\eta}{(\eta^2 + |x - y|^2)^{(n+1)/2}} dy. \tag{7}$$

This theorem[14] can be regarded as a converse of Theorem 3. In view of Theorem 4 the Dirichlet problem for $E^+_{n+1}$ always has precisely one solution in the following formulation, and this solution is given by the Poisson integral.

THE DIRICHLET PROBLEM. For a given function $f \in L^1(E_n)$ find a function $u(x, \eta)$ that is harmonic in $E^+_{n+1}$ and satisfies the following conditions:

a) $\lim_{\eta \searrow 0} u(x, \eta) = f(x)$ at all Lebesgue points of $f$.

b) $\int_{E_n} |u(x, \eta)| \cdot dx \le M < \infty$ for each $\eta > 0$.

c) $\lim_{\eta', \eta'' \searrow 0} \int_{E_n} |u(x, \eta') - u(x, \eta'')| \, dx = 0$.

**5.** PROBLEMS. 1. Prove that the Weierstrass and Poisson kernels have the semigroup property:

$$W(x, \tau_1 + \tau_2) = \int_{E_n} W(y, \tau_1) W(x - y, \tau_2) \, dy \qquad (\tau_1, \tau_2 > 0),$$

$$P(x, \eta_1 + \eta_2) = \int_{E_n} P(y, \eta_1) \cdot P(x - y, \eta_2) \, dy \qquad (\eta_1, \eta_2 > 0).$$

2. Prove that

$$\frac{1}{\sqrt{2\pi}} \int_{-\infty}^{\infty} \frac{1}{\cosh(x\sqrt{\pi/2})} e^{-itx} \, dx = \frac{1}{\cosh(t\sqrt{\pi/2})}.$$

3. Let $f \in L^1_{\text{loc}}(-\infty, \infty)$, i.e., $f \in L^1(a, b)$ for all finite $a$ and $b$, and suppose that

$$g(x) = \lim_{\substack{a \to -\infty \\ b \to \infty}} \frac{1}{\sqrt{2\pi}} \int_a^b f(u) e^{-iux} \, du$$

uniformly in each finite interval. Prove that

$$f(x) = \lim_{\lambda \to \infty} \frac{1}{\sqrt{2\pi}} \int_{-\lambda}^{\lambda} \left(1 - \frac{|u|}{\lambda}\right) g(u) e^{iux} \, du$$

almost everywhere and at all points of continuity of $f(x)$.

4. Let $f(x) = (\sin x/x)^r$, where $r$ is a positive integer. Prove that $\hat{f}(t) = 0$ for $|t| \ge r$.

---

(14) A proof is presented in Chapter 4 after the necessary information from the theory of harmonic functions has been given.

5. Prove that

$$\frac{12}{\pi}\int_{-\infty}^{\infty}\left(\frac{\sin(t/2)}{t}\right)^4 e^{-itx}\,dt = \begin{cases} 1-\dfrac{3x^2}{2}+\dfrac{3|x|^3}{4} & (|x|\le 1),\\ \dfrac{1}{4}(2-|x|)^3 & (1\le|x|\le 2),\\ 0 & (|x|\ge 2).\end{cases}$$

The function

$$I(z,n)=\frac{12}{\pi}\frac{\sin^4(nz/2)}{n^3z^4}$$

is called the *Jackson–Vallée-Poussin kernel.*

6. Prove that

$$\int_0^{\infty}\frac{e^{-itx}\,dx}{x^{\mu}}=\frac{\Gamma(1-\mu)}{it^{1-\mu}}\cdot e^{i\pi\mu/2}\qquad (0<\mu<1;\ t>0),$$

$$\int_0^{\infty}\frac{e^{-ax}-e^{-bx}}{x}e^{-itx}\,dx=\ln\frac{b+it}{a+it}\qquad (a>0,\ b>0).$$

7. Let

$$f(x)=\begin{cases}\cos^{p-1}x & (|x|<\pi/2),\\ 0 & (|x|>\pi/2),\end{cases}$$

where $p>0$. Prove that

$$\hat f(t)=\frac{\sqrt{\pi}}{2^{p-1/2}}\cdot\frac{\Gamma(p)}{\Gamma((p+1+t)/2)\cdot\Gamma((p+1-t)/2)}.$$

Prove that $\hat f(t)$ belongs to $L^1(-\infty,\infty)$ for $p>1$.

8. Prove that

$$\int_0^{\infty}e^{-t^2/2}\sin tz\cdot dt=e^{-z^2/2}\int_0^z e^{u^2/2}\,du.$$

9. Prove that

$$\sqrt{\frac{2}{\pi}}\int_0^{\infty}e^{-1/x}e^{-itx}\frac{dx}{\sqrt{x}}=\frac{1-i}{\sqrt{t}}e^{-\sqrt{2t}(1+i)}\qquad (t>0).$$

10. If $f\in L^1(-\infty,\infty)$ and

$$\lim_{a\to\infty}\int_{-\infty}^{\infty}\left|f(x)-\frac{1}{\sqrt{2\pi}}\int_{-a}^{a}\hat f(t)e^{itx}\,dt\right|dx=0,$$

then $\hat f$ has compact support (prove this).

CHAPTER 4

# Harmonic Functions. The Dirichlet Problem for a Ball and a Half-space

**1.** In this chapter we consider some properties of harmonic functions. Recall that the definition of harmonic functions was given in §4 of Chapter 3. There it was convenient to take the dimension of the space to be $n+1$. Here we denote it by $m$. However, $m$ will equal $n+1$ when we consider harmonic functions of a half-space, and $E_{n+1}^{+}$ will be the half-space.

We begin with an important averaging operator which we shall need below. Let

$$\theta(x) = \theta(|x|) = \begin{cases} C_m \cdot \exp \dfrac{1}{|x|^2 - 1} & (|x| < 1), \\ 0 & (|x| \geq 1), \end{cases}$$

where $x \in E_m$, while the coefficient $C_m$ is determined from the condition that

$$\int_{E_m} \theta(x)\, dx = \int_{K_m^0(1)} \theta(x)\, dx = 1.$$

The function $\theta(x)$ (as is known and easily verified) has continuous derivatives of any order. If $f$ is an arbitrary locally integrable function on $E_m$, then the function([15])

$$f_\rho(x) = \int_{E_m} \frac{1}{\rho^m} \theta\left(\frac{x-y}{\rho}\right) f(y)\, dy = \int_{E_m} \theta(z) \cdot f(x - \rho z)\, dz,$$

where $\rho > 0$, is an averaging of $f$. It has the following properties:

1) $f_\rho(x)$ has compact support if $f(x)$ does.

2) $f_\rho(x)$ has derivatives of any order, and they can be obtained by differentiating under the integral sign.

3) If $f \in L^p(E_m)$ for some finite $p \geq 1$, then([16])

$$\|f_\rho - f\|_p \leq \int_{E_m} \theta(z) \left\{ \int_{E_m} |f(x - \rho z) - f(x)|^p\, dx \right\}^{1/p} dz,$$

([15])To distinguish this function from that introduced in Chaper 1, §2, we would have to use the awkward notation $f_\rho(x, \theta)$ here, but we do not do this, since $\theta$ is a fixed function.

([16])$\|\cdot\|_p$ denotes the norm in $L^p(E_m)$.

which implies that

$$\lim_{\rho\to 0}\|f_\rho - f\|_p = 0.$$

The last conclusion is based on the fact that every function in $L^p(E_m)$ is continuous in the metric of this space.

**2.** Undoubtedly, the most important property of harmonic functions is the mean value theorem, formulated as follows.

THEOREM 1. *If an open set $G \subset E_m$ in which a harmonic function $u(x)$ is defined contains the ball $K_m^x(\rho)$, then*

$$\frac{1}{|S_{m-1}(\rho)|}\int_{S_{m-1}^x(\rho)} u(y)\,d\sigma = u(x) \tag{1}$$

*and*

$$\frac{1}{|K_m(\rho)|}\int_{K_m^x(\rho)} u(y)\,dy = u(x). \tag{2}$$

We shall not dwell on the proof of this theorem. It is contained in all mathematical physics courses. It is presented for $E_2$ and $E_3$ in Chapters 27 and 28 of Goursat's course of mathematical analysis. We remark only that this theorem can be derived from the well-known Green's formulas. We prove that either of these assertions (1) and (2) is a consequence of the other. Suppose, for example, that we want to derive (1) from (2). For this we represent (2) in the form

$$\int_0^\rho dr \int_{S_{m-1}^x(r)} u(y)\,d\sigma = u(x)\cdot|K_m(\rho)|.$$

Differentiating both sides with respect to $\rho$, we get that

$$\int_{S_{m-1}^x(\rho)} u(y)\,d\sigma = u(x)\frac{d}{d\rho}|K_m(\rho)| = u(x)\cdot|S_{m-1}(\rho)|.$$

It is a similar matter to derive (2) from (1).

Theorem 1 admits a converse, namely,

THEOREM 2. *If a function $u(x)$ is continuous in an open domain $G \subset E_m$ and satisfies the mean value theorem for any ball $K_m^x(\rho)$ in the interior of $G$, then $u(x)$ is harmonic in $G$.*

PROOF. By (1),

$$u_\delta(x) = \int_{E_m}\frac{1}{\delta^m}\theta\left(\frac{x-y}{\delta}\right)u(y)\,dy = \int_0^\infty \frac{1}{\delta^m}\theta\left(\frac{\rho}{\delta}\right)d\rho\int_{S_{m-1}^x(\rho)} u(y)\,d\sigma$$
$$= u(x)\int_0^\infty \frac{1}{\delta^m}\theta\left(\frac{\rho}{\delta}\right)|S_{m-1}|\rho^{m-1}\,d\rho = u(x)\int_{E_m}\theta(y)\,dy = u(x)$$

for any point $x \in G$ and any sufficiently small $\delta > 0$. As we see, $u(x) = u_\delta(x)$ for any $x \in G$ and any sufficiently small $\delta > 0$, and thus $u(x)$ has derivatives of all orders. We now find from Taylor's formula that

$$\begin{aligned} u(x) &= \frac{1}{|S_{m-1}(\rho)|}\int_{S^x_{m-1}(\rho)} u(y)\,d\sigma \\ &= \frac{1}{|S_{m-1}(\rho)|}\left\{\int_{S^x_{m-1}(\rho)} u(x)\,d\sigma + \sum_{i=1}^{m}\int_{S^x_{m-1}(\rho)}(y_i - x_i)\frac{\partial u(x)}{\partial x_i}\,d\sigma \right. \\ &\qquad \left. + \frac{1}{2}\sum_{i,k=1}^{m}\int_{S^x_{m-1}(\rho)}(y_i - x_i)(y_k - x_k)\frac{\partial^2 u(x)}{\partial x_i \partial x_k}\,d\sigma\right\} \\ &\qquad + O(\rho^3). \end{aligned} \tag{$*$}$$

Observing that by symmetry

$$\int_{S^x_{m-1}(\rho)}(y_i - x_i)\,d\sigma = 0; \quad \int_{S^x_{m-1}(\rho)}(y_i - x_i)(y_k - x_k)\,d\sigma = 0 \qquad (i \neq k);$$

$$\begin{aligned}\int_{S^x_{m-1}(\rho)}(y_i - x_i)^2\,d\sigma &= \int_{S^x_{m-1}(\rho)}(y_k - x_k)^2\,d\sigma = \frac{1}{m}\int_{S^x_{m-1}(\rho)}|y - x|^2\,d\sigma \\ &= \frac{\rho^2}{m}\cdot|S_{m-1}(\rho)|,\end{aligned}$$

we rewrite $(*)$ in the form

$$0 = \frac{1}{2m}\sum_{i=1}^{m}\frac{\partial^2 u}{\partial x_i^2} + O(\rho),$$

which yields the function.

THEOREM 3 (the maximum-minimum principle). *Let $u(x)$ be a real harmonic function on an open connected set (a domain) $G \subset E_m$ and assume that $M = \sup_G u(x) < \infty$ (respectively, $m = \inf_G u(x) > -\infty$). Then the following alternative holds: Either $u(x)$ takes the value $M$ (respectively, $m$) only on the boundary of $G$, or $u(x)$ is constant everywhere in $G$.*

PROOF. Suppose that $u(x)$ takes the value $M$ at some point $\dot{x} \in G$. By the mean value theorem, there is a ball $K^{\dot{x}}_m(\rho) \subset G$ at each point of which $u$ cannot be less than $M$, and hence $u$ equals $M$ everywhere in $K^{\dot{x}}_m(\rho)$. In other words, the set $\mathfrak{M}_M \subset G$ on which $u(x) = M$ is open in $G$. But since $u(x)$ is continuous, this set is also closed in $G$. Thus $\mathfrak{M}_M$ is both open and closed, and is hence either empty or the whole of $G$; but the first is excluded because $u(\dot{x}) = M$.

THEOREM 4 (PICARD-LIOUVILLE). *If a real function $u(x)$ is harmonic everywhere in $E_m$ and is semibounded, then it is constant.*

PROOF. For definiteness assume that $u(x)$ is semibounded above: $u(x) \leq M < \infty$. We take two points $x'$ and $x''$ and construct balls $K_m^{x'}(R')$ and $K_M^{x''}(R'')$ with radii such that the sphere $S_{m-1}^{x''}(R'')$ is tangent to $(S_{m-1}^{x'}(R')$ from the inside. Then, by (2),

$$(R')^m u(x') - (R'')^m u(x'') = \frac{1}{|K_m|} \cdot \int_{K_m^{x'}(R') \setminus K_m^{x''}(R'')} u(y)\,dy$$
$$\leq \frac{M}{|K_m|} \cdot \int_{K_m^{x'}(R') \setminus K_m^{x''}(R'')} dy = M(R'^m - R''^m),$$

which implies that

$$u(x') - \left(\frac{R''}{R'}\right)^m u(x'') \leq M \cdot \left[1 - \left(\frac{R''}{R'}\right)^m\right].$$

Since $R' = R'' + |x'' - x'|$, we obtain the inequality $u(x') - u(x'') \leq 0$ in the limit as $R' \to \infty$. It is proved similarly that $u(x'') - u(x') \leq 0$.

**3.** In considering harmonic functions in a ball domain one introduces a certain kernel, also bearing the name Poisson, used to express solutions of boundary value problems, first and foremost the Dirichlet problem. As a ball we take $K_m^0(1) = K_m$ and write the kernel (without normalization for the time being) in the form

$$h(x, \eta) = \frac{1 - r^2}{[1 - 2(x, \eta) + r^2]^{m/2}}.$$

Here $r = |x| < 1$ and $|\eta| = 1$. It can be verified by simple differentiation that $H(x, \eta)$ is harmonic with respect to $x$.

The verification is as follows. Let $D = 1 - 2(x, \eta) + r^2$, so that

$$D^{m/2} \cdot H = 1 - r^2. \qquad (**)$$

Differentiating with respect to $x_k$, we get

$$D^{m/2} \frac{\partial H}{\partial x_k} + mD^{m/2-1} \cdot (x_k - \eta_k) H = -2x_k,$$

$$D^{m/2} \frac{\partial^2 H}{\partial x_k^2} + 2mD^{m/2-1} \cdot (x_k - \eta_k) \frac{\partial H}{\partial x_k}$$
$$+ m(m-2) D^{m/2-2} (x_k - \eta_k)^2 H + mD^{m/2-1} \cdot H = -2.$$

Elimination of $\partial H / \partial x_k$ from these equalities gives us

$$D^{m/2+1} \cdot \frac{\partial^2 H}{\partial x_k^2} - m(m+2)(x_k - \eta_k)^2 D^{m/2-1} \cdot H + mD^{m/2} \cdot H$$
$$= 4m(x_k - \eta_k) x_k - 2D.$$

Summing $k$ from 1 to $m$, we get

$$D^{m/2+1} \cdot \Delta H - 2mD^{m/2} \cdot H + 2m(1-r^2) = 0,$$

and it remains to consider $(**)$.

The *Poisson integral for* $K_m$ has the form

$$u(x) = \frac{1}{|S_{m-1}|} \int_{S_{m-1}} \frac{1-|x|^2}{[1-2(x,y)+|x|^2]^{m/2}} f(y)\, d\sigma,$$

where $f$ is a given function on $S_{m-1}$, which we assume to be continuous. If $x = |x| \cdot \xi$ and $y = |y| \cdot \eta$, then the Poisson integral can be rewritten in the form

$$u(x) = u(r\xi) = \frac{1}{|S_{m-1}|} \int_{S_{m-1}} \frac{1-r^2}{[1-2r(\xi,\eta)+r^2]^{m/2}} f(\eta)\, d\sigma_\eta \qquad (r<1).$$

We now show that the function $H(x,\eta)/|S_{m-1}|$ satisfies the normalization condition (it is called the *normalized Poisson kernel*), i.e.,

$$\frac{1}{|S_{m-1}|} \int_{S_{m-1}} \frac{1-r^2}{[1-2r(\xi,\eta)+r^2]^{m/2}}\, d\sigma_\eta = 1.$$

For this, observe that $H(x,\eta) = H(r\xi,\eta) = H(r\eta,\xi)$. Since $H(x,\eta)$ is harmonic with respect to $x$, $|x|<1$, the mean value theorem implies that

$$\begin{aligned} 1 = H(0,\eta) &= \frac{1}{|S_{m-1}(r)|} \int_{S_{m-1}(r)} H(x,\eta)\, d\sigma_x \\ &= \frac{1}{|S_{m-1}|} \int_{S_{m-1}} H(r\xi,\eta)\, d\sigma_\xi. \end{aligned}$$

Therefore, by the symmetry of $H(x,\eta)$,

$$\frac{1}{|S_{m-1}|} \int_{S_{m-1}} H(r\eta,\xi)\, d\sigma_\xi = 1.$$

Replacing $\xi$ by $\eta$ and $\eta$ by $\xi$, we get that

$$\frac{1}{|S_{m-1}|} \int_{S_{m-1}} H(r\xi,\eta)\, d\sigma_\eta = 1.$$

The following theorem gives a solution of the classical Dirichlet problem for a ball.

THEOREM 5. *For any continuous function $f$ on $S_{m-1}$ there exists one and only one function $u(x)$ harmonic in $K_m$ and continuous in $\overline{K}_m = K_m \cup S_{m-1}$ that coincides with $f(x)$ on $S_{m-1}$. This function can be represented by the Poisson integral* $(r = |x| < 1)$

$$u(x) = \frac{1}{|S_{m-1}|} \int_{S_{m-1}} \frac{(1-r^2)\cdot f(\eta)}{[1-2(x,\eta)+r^2]^{m/2}}\, d\sigma. \tag{3}$$

PROOF. It follows at once from the maximum and minimum principle (Theorem 3) that the problem has at most one solution. Since (3) implies that $u(x)$ is harmonic interior to $K_m$, it is only necessary to prove that $u(r\xi)$ tends to $f(\xi)$ uniformly on $S_{m-1}$ as $r \to 1$. With this goal we consider the difference

$$u(r\xi) - f(\xi) = \frac{1}{|S_{m-1}|} \int_{S_{m-1}} \frac{(1-r^2)[f(\eta) - f(\xi)]}{[1 - 2r(\xi, \eta) + r^2]^{m/2}} \, d\sigma_\eta. \tag{4}$$

If we set $(\xi, \eta) = \cos\gamma$, then $\gamma$ $(0 \le \gamma \le \pi)$ will be the angle between the vectors going out from the center of $S_{m-1}$ to its points $\xi$ and $\eta$. The continuity of $f$ on $S_{m-1}$ means that for any $\varepsilon > 0$ there is a $\delta$ such that $|f(\eta) - f(\xi)| < \varepsilon$ if $\gamma < \delta$. Having made this remark, we represent (4) in the form

$$\begin{aligned} u(r\xi) - f(\xi) &= \frac{1}{|S_{m-1}|} \int_{S'_{m-1}} \frac{(1-r^2)[f(\eta) - f(\xi)]}{[1 - 2r(\xi, \eta) + r^2]^{m/2}} \, d\sigma_\eta \\ &\quad + \frac{1}{|S_{m-1}|} \cdot \int_{S''_{m-1}} \cdots, \end{aligned}$$

where $S'_{m-1}$ is the part of $S_{m-1}$ on which $0 \le \gamma < \delta$, and $S''_{m-1}$ is its complement, on which $\delta \le \gamma \le \pi$. Then for any $r$ $(0 \le r < 1)$

$$\frac{1}{|S_{m-1}|} \cdot \left| \int_{S'_{m-1}} \cdots \right| \le \frac{\varepsilon}{|S_{m-1}|} \int_{S_{m-1}} \frac{(1-r^2)\, d\sigma_\eta}{[1 - 2r\cos\gamma + r^2]^{m/2}} = \varepsilon.$$

On the other hand,

$$\begin{aligned} \frac{1}{|S_{m-1}|} \cdot \left| \int_{S''_{m-1}} \cdots \right| &\le \frac{2M}{|S_{m-1}|} \int_{S''_{m-1}} \frac{1 - r^2}{[(1-r)^2 + 4r\sin^2(\gamma/2)]^{m/2}} \, d\sigma_\eta \\ &\le \frac{2M \cdot (1-r^2)}{(4r\sin^2(\delta/2))^{m/2}} \frac{1}{|S_{m-1}|} \int_{S''_{m-1}} d\sigma \le \frac{2M \cdot (1-r^2)}{(4r\sin^2(\delta/2))^{m/2}}, \end{aligned}$$

where $M = \max|f(\xi)|$. The right-hand side of the resulting inequality is less than $\varepsilon$ for the fixed $\delta$ if $r > 1 - \vartheta$ and $\vartheta > 0$ is sufficiently small. Thus, for any $\varepsilon > 0$ and any $\xi \in S_{m-1}$ we have that $|u(r\xi) - f(\xi)| < 2\varepsilon$ if $r > 1 - \vartheta$. The theorem is proved.

An important consequence of Theorem 5 is the so-called *symmetry principle*, expressed as the following theorem.

THEOREM 6. *Suppose that $u(x)$ is harmonic in a domain $G$ lying above the plane $x_m = 0$ and having part of this plane as a piece of its boundary. If $u(x)$ is continuous in $\overline{G}$ and equal to 0 on the planar boundary of $G$, then $u(x)$ can be extended harmonically to the domain $G'$ symmetric to $G$ with* *respect to the plane $x_m = 0$ by setting $u(x') = -u(x)$ when $x'$ and $x$ are symmetric with respect to this plane.*

PROOF. It suffices to consider the case when $G$ is a half-ball with center on the plane $x_m = 0$, and it can clearly be assumed for simplicity that the radius of the ball is equal to 1. We take the Poisson integral

$$v(x) = \frac{1}{|S_{m-1}|} \int_{S_{m-1}} \varphi(\eta) \frac{1 - |x|^2}{[1 - 2(x, \eta) + |x|^2]^{m/2}} \, d\sigma,$$

where $\varphi(\eta) = u(\eta)$ $(\eta_m \geq 0)$ and $\varphi(\eta) = -u(\eta')$ $(\eta_m \leq 0)$. By Theorem 5, $v(x)$ is harmonic in $K_m$ and continuous in its closure, and it coincides with $u(x)$ on the upper half of $S_{m-1}$. Since $(x, \eta) = x_1\eta_1 + \cdots + x_{m-1}\eta_{m-1} = (x, \eta')$ for $x_m = 0$, while, on the other hand, $\varphi(\eta) = -\varphi(\eta')$, the function $v(x)$ vanishes for $x_m = 0$, like $u(x)$. Hence, the equality $v(x) = u(x)$ holds on the whole boundary of the upper half-ball, and thus at interior points of this half-ball. It can be proved similarly that $v(x) = -u(x')$ at the points of the lower half-ball. The theorem is proved.

**4.** Let us now turn to the Dirichlet problem for $E_{n+1}^+$, which we have already studied in §4 of Chapter 3; in particular, we prove Theorem 4 of Chapter 3, which was only stated there. We begin with a certain variant of Theorem 3 in Chapter 3.

THEOREM 7. *Let $f$ be a continuous bounded function on $E_n$, and let*

$$u(x, \eta) = \frac{2}{|S_n|} \int_{E_n} f(y) \frac{\eta \, dy}{(\eta^2 + |y - x|^2)^{(n+1)/2}}$$

*be its Poisson integral on $E_{n+1}^+$. Then the harmonic function $u(x, \eta)$ $(\eta > 0)$ has the following properties:*

1) $|u(x, \eta)| \leq M$ *for any* $\eta > 0$, *where* $M = \sup_{E_n} |f|$.

2) $u(x, \eta) \rightrightarrows f(x)$ *on each compact set* $Q \subset E_n$ *as* $\eta \searrow 0$.

PROOF. Assertion 1) is a consequence of the normalization of the Poisson kernel. To prove 2) we write the equality

$$\begin{aligned} u(x, \eta) - f(x) &= \frac{2}{|S_n|} \int_{|z| \leq R} [f(x + \eta z) - f(x)] \frac{dz}{(1 + |z|^2)^{(n+1)/2}} \\ &\quad + \frac{2}{|S_n|} \int_{|z| > R} [f(x + \eta z) - f(x)] \frac{dz}{(1 + |z|^2)^{(n+1)/2}} \\ &= I_1 + I_2. \end{aligned}$$

Here we have set

$$|I_2| \leq \frac{4M}{|S_n|} \int_{|z| > R} \frac{dz}{(1 + |z|^2)^{(n+1)/2}} \leq \frac{4M}{R}$$

for $\eta > 0$, $x \in E_n$, and $R > 1$. On the other hand, for each $R$ we can choose an $\eta_0$ such that $|f(x+\eta z) - f(x)| < \varepsilon$ for all $x \in Q$ if $|z| < R$ and $\eta \leq \eta_0$. This implies 2).

To prove Theorem 4 of Chapter 3 we need a lemma.

LEMMA. *If the function $u(x, \eta)$ is harmonic in $E_{n+1}$ and is bounded in the half-space $\eta \geq \eta_0$ for any $\eta_0 > 0$, then for any $\eta_1 > 0$ and $\eta_2 > 0$*

$$u(x, \eta_1 + \eta_2) = \int_{E_n} u(t, \eta_1) P(x - t, \eta_2)\, dt.$$

PROOF. We fix $\eta_0 > 0$ and let

$$w(x, \eta) = u(x, \eta + \eta_0)$$
$$w_1(x, \eta) = \int_{E_n} u(t, \eta_0) P(x - t, \eta)\, dt \qquad (\eta > 0).$$

Since $u(x, \eta_0)$ is continuous, $w_1(x, \eta) \rightrightarrows u(x, \eta_0)$ on each compact subset of $E_n$ as $\eta \to 0$, by Theorem 7. The same property is clearly enjoyed by $w(x, \eta)$. Therefore, the difference

$$v(x, \eta) = w(x, \eta) - w_1(x, \eta)$$

is harmonic in $E_{n+1}^+$, continuous in the closure $\overline{E_{n+1}^+}$, and equal to zero on the boundary of $E_{n+1}^+$, i.e., on the hyperplane $\eta = 0$. Moreover, this function is bounded by a condition of the lemma. Extending $v(x, \eta)$ by the symmetry principle, we get a bounded harmonic function on the whole space $E_{n+1}$. By the Picard-Liouville theorem, this function must be constant, i.e., $v(x, \eta) = \text{const}$, and hence $v(x, \eta) = 0$, since this is true for $\eta = 0$. Therefore,

$$u(x, \eta + \eta_0) = \int_{E_n} u(t, \eta_0) P(x - t, \eta)\, dt, \tag{5}$$

and the lemma is proved.

PROOF OF THEOREM 4 IN CHAPTER 3. We first prove, using only b), that $u(x, \eta)$ is bounded in the half-space $\eta \geq \eta_0$, for any $\eta_0 > 0$. The mean value theorem in the form (2) suffices for this. It gives us that for any $\eta > 0$

$$\begin{aligned}
|u(x, \eta)| &\leq \frac{1}{|K_{n+1}(\eta/2)|} \int_{K_{n+1}^{(x,\eta)}(\eta/2)} |u(y, \zeta)|\, dy\, d\zeta \\
&\leq \frac{1}{|K_{n+1}(\eta/2)} \int_{\eta/2}^{3\eta/2} d\zeta \int_{E_n} |u(y, \zeta)|\, dy \\
&\leq \frac{M\eta}{|K_{n+1}(\eta/2)|} = \frac{2^{n+1} M}{|K_{n+1}(1)| \eta^n}.
\end{aligned}$$

It is proved that $u(x, \eta)$ is bounded in each half-space $\eta \geq \eta_0 > 0$. Consequently, the lemma is applicable, and formula (5) holds for any $\eta_0 > 0$ and $\eta > 0$ due to condition b) alone. We now turn to condition c*). It expresses the fact that the "continuous sequence" of elements $u(x, \eta_0)$ of $L^1(E_n)$ converges in the Cauchy sense as $\eta_0 \searrow 0$. By the completeness of $L^1(E_n)$, there exists a function $f \in L^1(E_n)$ such that

$$\lim_{\eta_0 \searrow 0} \int_{E_n} |u(t, \eta_0) - f(t)|\, dt = 0.$$

With the help of the function $f$ we rewrite (5) in the form

$$\begin{aligned} u(x, \eta_0 + \eta) &- \int_{E_n} f(t) P(x - t, \eta)\, dt \\ &= \int_{E_n} \{u(t, \eta_0) - f(t)\} P(x - t, \eta)\, dt, \end{aligned} \tag{6}$$

where $\eta$ is a fixed positive number. Since

$$\left| \int_{E_n} \{u(t, \eta_0) - f(t)\} P(x - t, \eta)\, dt \right| \leq \frac{1}{|S_n|} \cdot \frac{1}{\eta^n} \int_{E_n} |u(t, \eta_0) - f(t)|\, dt,$$

the right-hand side of (6) tends to zero uniformly with respect to $x$ as $\eta_0 \searrow 0$. Consequently, (6) implies that

$$u(x, \eta) - \int_{E_n} f(t) P(x - t, \eta)\, dt = 0,$$

where $x \in E_n$ and $\eta > 0$ are arbitrary. Thus, the theorem is proved.

CHAPTER 5

# The Fourier Transform in $L^2$

**1.** Let us proceed to the definition and study of the Fourier transform in $L^2(E_n)$. With this purpose we first introduce the space $L^{1,2}(E_n) = L^1(E_n) \cap L^2(E_n)$. If $f$ belongs to this space, then $\hat{f}$ is defined, since $f \in L^1(E_n)$. But since $f$ is also in $L^2(E_n)$ the function $\hat{f}$ has an essential additional property, which we now derive. Consider the function

$$g(x) = \int_{E_n} f(x+t)\overline{f(t)}\,dt, \tag{1}$$

which is the convolution of two $L^1(E_n)$-functions, namely,

$$g(x) = \int_{E_n} f(x-t)\overline{f(-t)}\,dt.$$

Therefore, $g(x) \in L^1(E_n)$. By the Cauchy-Schwarz-Bunyakovskiĭ inequality,

$$|g(x)| \leq \int_{E_n} |f(t)|^2\,dt = \|f\|_2^2, \tag{$\alpha$}$$

i.e., $g$ is bounded in $E_n$. Further, again by the Cauchy-Schwarz-Bunyakovskiĭ inequality,

$$|g(x'') - g(x')| \leq \|f\|_2 \cdot \sqrt{\int_{E_n} |f(t+x'') - f(t+x')|^2\,dt},$$

and since every function in $L^p(E_n)$ ($p \geq 1$) is continuous in the metric of this space, $g(x)$ is uniformly continuous in $E_n$. If we write $h(t) = \overline{f(-t)}$ for the time being, we find from (1) that

$$\hat{g}(x) = (\sqrt{2\pi})^n \hat{f}(x)\hat{h}(x),$$

and, on the other hand,

$$\begin{aligned}\hat{h}(x) &= \frac{1}{(\sqrt{2\pi})^n}\int_{E_n} \overline{f(-t)}e^{-i(t,x)}\,dt \\ &= \frac{1}{(\sqrt{2\pi})^n}\int_{E_n} \overline{f(u)e^{-i(u,x)}}\,du = \overline{\hat{f}(x)}.\end{aligned}$$

Consequently,

$$\hat{g}(x) = (\sqrt{2\pi})^n |\hat{f}(x)|^2. \tag{$\beta$}$$

Accordingly, $g(x)$ is bounded because of $(\alpha)$, and $\hat{g}(x) \geq 0$ because of $(\beta)$. Therefore, $\hat{g} \in L^1(E_n)$ by Theorem 2 of Chapter 3, and thus, by the inversion formula,

$$g(x) = \frac{1}{(\sqrt{2\pi})^n} \int_{E_n} \hat{g}(t) e^{i(t,x)}\, dt$$

for almost all $x$, hence for all $x$, since $g(x)$ is continuous. From this, by (1) and $(\beta)$,

$$\int_{E_n} f(x+t)\overline{f(t)}\, dt = \int_{E_n} |\hat{f}(t)|^2 e^{i(t,x)}\, dt.$$

In particular, for $x = 0$

$$\int_{E_n} |f(t)|^2\, dt = \int_{E_n} |\hat{f}(t)|^2\, dt. \tag{2}$$

Thus, if $f \in L^{1,2}(E_n)$, then $\hat{f} \in L^2(E_n)$, and the *Parseval equality* (2) is valid. By the usual device, we conclude from this equality that for any two functions $f$, $F \in L^{1,2}(E_n)$

$$\int_{E_n} f(t)\overline{F(t)}\, dt = \int_{E_n} \hat{f}(t)\overline{\hat{F}(t)}\, dt. \tag{3}$$

This is the additional property mentioned above.

**2.** We can now give the definition of the Fourier transform of an arbitrary function in $L^2(E_n)$. Let $\{f_m\}_1^\infty \subset L^1(E_n) \cap L^2(E_n)$ be a sequence such that

$$\|f_m - f\|_2 \to 0 \qquad (m \to \infty).$$

Such a sequence always exists; in other words, $L^{1,2}(E_n)$ is dense in $L^2(E_n)$. For example, $f_m$ can be taken to be

$$f_m(x) = \begin{cases} f(x) & (-m/2 < x_k < m/2; k = 1, 2, \ldots, n), \\ 0 & \text{outside this cube.} \end{cases} \tag{4}$$

Having chosen $\{f_m\}_1^\infty$, we note that $\hat{f}_m$ exists for each $m$, and

$$\|\hat{f}_m\|_2 = \|f_m\|_2, \qquad \|\hat{f}_{m'} - \hat{f}_m\|_2 = \|f_{m'} - f_m\|_2.$$

Since $\|f_m - f\|_2 \to 0$ $(m \to \infty)$, it follows that $\|f_{m'} - f_m\| \to 0$ $(m, m' \to \infty)$, and hence $\|\hat{f}_{m'} - \hat{f}_m\|_2 \to 0$ $(m, m' \to \infty)$, i.e., the sequence $\{\hat{f}_m\}_1^\infty$ converges in the Cauchy sense. Since $L^2(E_n)$ is complete, there exists a function $\hat{f} \in L^2(E_n)$ such that $\|\hat{f}_m - \hat{f}\|_2 \to 0$. This function, which is determined only to within values on a set of measure zero, is called the *limit in the mean* of the sequence $\hat{f}_m$, written $\hat{f}(x) = \text{l.i.m.}_{m\to\infty}\hat{f}_m(x)$, and is taken as the Fourier transform of $f$. The equality

$$\|\hat{f}\|_2 = \|f\|_2. \tag{5}$$

implies that this definition is unambiguous, i.e., the transform is independent (to within values on a set of measure zero) of the choice of the sequence $\{f_m\}_1^\infty \subset L^{1,2}(E_n)$ converging to $f$ in $L^2(E_n)$.

If the sequence $\{f_m\}_1^\infty$ is defined by (4), then we get the expression

$$\hat{f}(x) = \underset{m\to\infty}{\text{l. i. m.}} \frac{1}{(\sqrt{2\pi})^n} \int_{Q_n(m)} f(t)e^{-i(t,x)}\,dt$$

for the Fourier transform $\hat{f}$, where $Q_n(m)$ is the cube $-m/2 < x_k < m/2$ $(k = 1, \dots, n)$ in $E_n$. Note further that, by linearity and (5), we have the general Parseval equality

$$\int_{E_n} f(x)\overline{g(x)}\,dx = \int_{E_n} \hat{f}(x)\overline{\hat{g}(x)}\,dx.$$

It is not hard to see that, as in $L^1(E_n)$, the following simple rules are valid for the Fourier transform in $L^2(E_n)$:

1) If $F(x) = f(x + c)$, then $\hat{F}(x) = e^{i(c,x)}\hat{f}(x)$.
2) If $F(x) = e^{-i(c,x)}f(x)$, then $\hat{F}(x) = \hat{f}(x + c)$.
3) If $F(x) = f(Rx)$, $R \neq 0$ a real number, then $\hat{F}(x) = \hat{f}(x/R)/|R|^n$.

**3.** As an example, we find the Fourier transform of the function

$$\Psi_a(t) = \frac{1}{(\sqrt{2\pi})^n} \prod_{k=1}^n \frac{\sin at_k}{at_k} \in L^2(E_n),$$

where $a$ is a positive parameter. By what we have proved,

$$\hat{\Psi}_a(x) = \underset{m\to\infty}{\text{l. i. m.}} \frac{1}{(\sqrt{2\pi})^n} \int_{Q_n(m)} \Psi_a(t)e^{-i(t,x)}\,dt.$$

For fixed $x \in E_n$ the limit

$$\lim_{m\to\infty} \frac{1}{(\sqrt{2\pi})^n} \int_{Q_n(m)} \Psi_a(t)e^{-i(t,x)}\,dt$$

exists and is easily computed. Indeed,

$$\int_{Q_n(m)} \Psi_a(t)e^{-i(t,x)}\,dt = \frac{1}{(\sqrt{2\pi})^n} \prod_{k=1}^n \int_{-m/2}^{m/2} \frac{\sin at_k}{at_k} e^{-it_kx_k}\,dt_k,$$

and since

$$\begin{aligned}\lim_{m\to\infty}\frac{1}{2\pi}\int_{-m/2}^{m/2}\frac{\sin a\tau}{a\tau}e^{-i\tau\xi}\,d\tau &= \lim_{m\to\infty}\frac{1}{\pi}\int_0^{m/2}\frac{\sin a\tau}{a\tau}\cos\tau\xi\,d\tau\\ &= \frac{1}{\pi a}\int_0^{\infty}\frac{\sin(a+\xi)\tau+\sin(a-\xi)\tau}{2\tau}\,d\tau\\ &= \frac{1}{4a}\{\operatorname{sgn}(a+\xi)+\operatorname{sgn}(a-\xi)\}\\ &= \begin{cases}1/2a, & |\xi|<a,\\ 0, & |\xi|>a,\\ 1/4a, & |\xi|=a,\end{cases}\end{aligned}$$

it follows that

$$\lim_{m\to\infty}\frac{1}{(\sqrt{2\pi})^n}\int_{Q_n(m)}\Psi_a(t)e^{-i(t,x)}\,dt=\frac{1}{(2a)^n}$$

for $x\in Q_n(2a)$, and this limit is equal to zero for $x\notin\overline{Q_n(2a)}$.

As we know, the pointwise limit and the $L^p$-limit of an $L^p$-sequence ($p\geq 1$) coincide almost everywhere if both limits exist.([17]) Therefore, denoting by $Q_a(x)$ ($x\in E_n$) the characteristic function of the cube $Q_n(2a)$ divided by its volume, we get from the preceding computations that $\hat{\Psi}_a(x)=\Omega_a(x)$ (a.e.). It is easy to verify directly that $\hat{\Omega}_a(x)=\Psi_a(x)$ (a.e.) also.

**4.** We now show that the Fourier transform on $L^2(E_n)$ can be represented in another important form, and in passing we also obtain an inversion formula. With this goal, we take an $f\in L^2(E_n)$ and its Fourier transform $\hat{f}$, and write the following two Parseval equalities:

$$\begin{aligned}\int_{E_n}f(t)\overline{e^{i(x,t)}\Psi_a(t)}\,dt &= \int_{E_n}\hat{f}(t)\overline{\Omega_a(t-x)}\,dt,\\ \int_{E_n}f(t)\overline{\Omega_a(t-x)}\,dt &= \int_{E_n}\hat{f}(t)\overline{e^{-i(t,x)}\Psi_a(t)}\,dt,\end{aligned}$$

where we have used rules 1) and 2) given at the end of §2. Substituting the explicit expressions for $\Omega_a$ and $\Psi_a$ in these equalities, we reduce them to the form

$$\begin{aligned}\frac{1}{(\sqrt{2\pi})^n}\int_{E_n}f(t)e^{-i(t,x)}\prod_{k=1}^{n}\frac{\sin at_k}{at_k}\,dt &= \frac{1}{|Q_n(2a)|}\int_{Q_n(2a)}\hat{f}(x+t)\,dt,\\ \frac{1}{(\sqrt{2\pi})^n}\int_{E_n}\hat{f}(t)e^{i(t,x)}\prod_{k=1}^{n}\frac{\sin at_k}{at_k}\,dt &= \frac{1}{|Q_n(2a)|}\int_{Q_n(2a)}f(x+t)\,dt.\end{aligned}\tag{6}$$

([17])We recall that this proposition is obtained as a byproduct in the process of proving that $L^p$ is complete.

This gives us that for almost all $x \in E_n$

$$\hat{f}(x) = \lim_{a\to 0} \frac{1}{(\sqrt{2\pi})^n} \int_{E_n} f(t) e^{-i(t,x)} \prod_{k=1}^{n} \frac{\sin at_k}{at_k}\, dt,$$
$$f(x) = \lim_{a\to 0} \frac{1}{(\sqrt{2\pi})^n} \int_{E_n} \hat{f}(t) e^{i(t,x)} \prod_{k=1}^{n} \frac{\sin at_k}{at_k}\, dt. \tag{7}$$

The first formula shows that the Fourier transform can be represented on $L^2(E_n)$ not only as a limit in the mean (l.i.m.), but also as a pointwise limit almost everywhere. The original function is determined similarly from its transform. Further, the formulas (7) show that every $L^2(E_n)$-function is the transform of some $L^2(E_n)$-function, and this together with the Parseval equality means that *the Fourier transform is a unitary operator on* $L^2(E_n)$.

We dwell in detail on the case $n = 1$. In this case the second of the formulas (6) has the form

$$\frac{1}{\sqrt{2\pi}} \int_{-\infty}^{\infty} \hat{f}(t) e^{itx} \frac{\sin at}{at}\, dt = \frac{1}{2a} \int_{-a}^{a} f(x+t)\, dt.$$

Replacing $x$ by $u$ and integrating with respect to $u$ from 0 to $x$, we find that

$$\frac{1}{\sqrt{2\pi}} \int_{-\infty}^{\infty} \hat{f}(t) \cdot \frac{e^{itx} - 1}{it} \cdot \frac{\sin at}{at}\, dt = \frac{1}{2a} \int_{-a}^{a} dt \int_{a}^{x} f(u+t)\, du.$$

Passing to the limit as $a \to 0$, we get that

$$\frac{1}{\sqrt{2\pi}} \int_{-\infty}^{\infty} \hat{f}(t) \frac{e^{itx} - 1}{it}\, dt = \int_{0}^{x} f(u)\, du.$$

Hence,

$$f(x) = \frac{1}{\sqrt{2\pi}} \cdot \frac{d}{dx} \int_{-\infty}^{\infty} \hat{f}(t) \frac{e^{itx} - 1}{it}\, dt$$

almost everywhere. Similarly, the first of the formulas in (6) gives us

$$\hat{f}(x) = \frac{1}{\sqrt{2\pi}} \cdot \frac{d}{dx} \int_{-\infty}^{\infty} f(t) \frac{e^{-itx} - 1}{-it}\, dt. \tag{8}$$

The Fourier transform on $L^2$ was first constructed and studied by Plancherel, and it is therefore called the *Fourier-Plancherel transform.*

**5. Problems.** 1. Prove that relation (4) in Chapter 2 is valid also for $f, g \in L^2(E_n)$.

2. Prove that the convolution

$$F(x) = \int_{E_n} f(x-y) g(y)\, dy$$

exists for almost all $x \in E_n$ in each of the following cases (in addition to those considered above in Chapter 2, §2):

a) Both $f$ and $g$ are in $L^2(E_n)$.

b) One function is in $L^1(E_n)$, and the other in $L^2(E_n)$.

Prove that $F \in L^\infty(E_n)$ in case a), and $F \in L^2(E_n)$ in case b).

3. Let

$$F(x) = \int_{E_n} f(x-y)g(y)\,dy,$$

where $f \in L^2(E_n)$, and $g$ is in either $L^2(E_n)$ or $L^1(E_n)$. Prove that $\hat{F}(t) = (\sqrt{2\pi})^n \hat{f}(t)\hat{g}(t)$.

*Hint.* Let $g \in L^2$; prove that the Fourier-Plancherel transform of the function $\varphi(t) = \hat{f}(t)e^{i(x,t)}$ is equal to $\hat{\varphi}(y) = f(x-y)$, and use the equality

$$\int_{E_n} \hat{\varphi}(y)g(y)\,dy = \int_{E_n} \varphi(y)\hat{g}(y)\,dy.$$

Now let $g \in L^1$ and (for simplicity) let $n = 1$; setting $h = \hat{f} \cdot \hat{g}$ and noting that $h \in L^2$, prove the equality

$$\int_{-\infty}^{\infty} h(y)\frac{e^{ity}-1}{iy}\,dy = \frac{1}{\sqrt{2\pi}} \int_{-\infty}^{\infty} \int_{-\infty}^{\infty} g(u)\,du \int_{-\infty}^{\infty} \hat{f}(y) \left\{ \frac{\overline{e^{i(u-t)y} - e^{iuy}}}{-iy} \right\} dy.$$

Using the Fourier transform of the characteristic function $\chi(y)$ of the interval $[-u, -u+t]$, deduce from this that

$$\int_{-\infty}^{\infty} h(y)\frac{e^{ity}-1}{iy}\,dy = \int_0^t dv \int_{-\infty}^{\infty} f(v-u)g(u)\,du.$$

Then differentiate with respect to $t$ and use (8).

CHAPTER 6

# Hermite Functions

**1.** In this and the next chapter we consider how the Fourier transform is connected with certain special functions. We begin with the so-called Hermite functions and the Tchebycheff-Hermite polynomials. The *Hermite functions* are defined by

$$\varphi_k(t) = e^{t^2/2}\frac{d^k}{dt^k}e^{-t^2}, \tag{1}$$

$k = 0, 1, \ldots$. We easily prove by induction that

$$d^k e^{-t^2}/dt^k = (-1)^k e^{-t^2} H_k(t),$$

where $H_k(t)$ is a $k$th-degree polynomial with leading term $(2t)^k$, and that $H_k(t)$ is an even function for even $k$ and an odd function for odd $k$. The polynomials $H_k(t)$ are called the *Tchebycheff-Hermite polynomials.* It follows from (1) that $\varphi_k(t) = (-1)^k e^{-t^2/2} H_k(t)$.

The main property of the Hermite functions is their orthogonality:

$$\int_{-\infty}^{\infty} \varphi_k(t)\varphi_l(t)\,dt = 0 \qquad (k \neq l). \tag{2}$$

The relations (2) are clearly equivalent to the following:

$$\int_{-\infty}^{\infty} \varphi_k(t)e^{-t^2/2}t^l\,dt = 0 \qquad (l = 0, 1, \ldots, k-1).$$

But, by (1), these relations can be represented in the form

$$\int_{-\infty}^{\infty} t^l\left(\frac{d^k}{dt^k}e^{-t^2}\right)dt = 0 \qquad (l = 0, 1, \ldots, k-1)$$

and can be immediately verified by $k$-fold integration by parts.

We now prove that

$$\hat{\varphi}_k(t) = (-i)^k\varphi_k(t). \tag{3}$$

Indeed,

$$\begin{aligned}
\hat{\varphi}_k(t) &= \frac{1}{\sqrt{2\pi}} \int_{-\infty}^{\infty} e^{x^2/2} \frac{d^k e^{-x^2}}{dx^k} \cdot e^{-itx}\, dx \\
&= \frac{1}{\sqrt{2\pi}} \int_{-\infty}^{\infty} e^{(x-it)^2/2} e^{t^2/2} \frac{d^k e^{-x^2}}{dx^k}\, dx \\
&= \frac{(-1)^k}{\sqrt{2\pi}} e^{t^2/2} \int_{-\infty}^{\infty} e^{-x^2} \frac{d^k}{dx^k} e^{(x-it)^2/2}\, dx \\
&= \frac{(-i)^k}{\sqrt{2\pi}} e^{t^2/2} \frac{d^k}{dt^k} \int_{-\infty}^{\infty} e^{-x^2} e^{(x-it)^2/2}\, dx \\
&= \frac{(-i)^k}{\sqrt{2\pi}} e^{t^2/2} \frac{d^k}{dt^k} \left[ e^{-t^2/2} \int_{-\infty}^{\infty} e^{-x^2/2} e^{-itx}\, dx \right] = (-i)^k e^{t^2/2} \frac{d^k}{dt^k} e^{-t^2}.
\end{aligned}$$

Relation (3), which is true for $k = 0, 1, \dots$, means that the Hermite functions are eigenfunctions of the Fourier transform, and there are only four eigenvalues: $1, i, -1, -i$. Each of them thus has an infinite set of eigenfunctions.

**2. Lemma.** *If $f \in L^2(-\infty, \infty)$ and*

$$\int_{-\infty}^{\infty} f(t)\varphi_k(t)\, dt = 0 \qquad (k = 0, 1, \dots), \tag{$\alpha$}$$

*then $f(t) = 0$ almost everywhere.*

Proof. Instead of the conditions ($\alpha$) we can clearly assume that

$$\int_{-\infty}^{\infty} f(t) e^{-t^2/2} t^k\, dt = 0 \qquad (k = 0, 1, \dots). \tag{$\beta$}$$

Consider the function

$$F(z) = \frac{1}{\sqrt{2\pi}} \int_{-\infty}^{\infty} f(t) e^{-t^2/2} e^{-itz}\, dt,$$

where $z = x + iy$ is a complex variable.

Since $f(t)e^{-t^2/4} \in L^1(-\infty, \infty)$, while, on the other hand, $e^{-t^2/4}e^{-itz}t^k$ is a bounded function of $t$ on the whole axis for any complex number $z$ and any positive integer $k$, $F(z)$ is holomorphic in the whole $z$-plane, and all its derivatives

$$F^{(k)}(z) = \frac{(-i)^k}{\sqrt{2\pi}} \int_{-\infty}^{\infty} f(t) e^{-t^2/2} e^{-itz} t^k\, dt$$

are equal to 0 at $z = 0$ in view of ($\beta$). By the uniqueness theorem, $F(z) \equiv 0$. Hence,

$$0 = F(x) = \frac{1}{\sqrt{2\pi}} \int_{-\infty}^{\infty} f(t) e^{-t^2/2} e^{-itx}\, dt \qquad (-\infty < x < \infty).$$

Since the function $f(t)e^{-t^2/2}$ belongs to $L^1(-\infty, \infty)$, the lemma follows from the uniqueness theorem for the Fourier transform in $L^1$.

It follows from the lemma that *the set of Hermite functions is exhausted by the collection of all eigenfunctions of the Fourier operator, and the whole spectrum of this operator is exhausted by the points* $\pm 1$ *and* $\pm i$. Moreover, the lemma gives us that *each function in* $L^2(-\infty, \infty)$ *can be expanded in the* $L^2$*-metric in a generalized Fourier series of Hermite functions.*

**3.** The *normalizing coefficients*, i.e., the numbers

$$\alpha_k^2 = \int_{-\infty}^{\infty} [\varphi_k(t)]^2 \, dt,$$

are needed for series expansions in Hermite functions. It is not hard to show that $\alpha_k^2 = 2^k k! \sqrt{\pi}$. The *orthonormal Hermite system* $\{\varphi_k(t)/\alpha_k\}_0^\infty$ is obtained with the help of the numbers $\alpha_k$. For any function $f \in L^2(-\infty, \infty)$ the expansion has the form

$$f(t) \overset{L^2}{=} \sum_{k=0}^{\infty} \frac{c_k}{\alpha_k^2} \varphi_k(t), \tag{4}$$

where the generalized Fourier coefficients $c_k$ are given by

$$c_k = \langle f, \varphi_k \rangle = \int_{-\infty}^{\infty} f(t)\varphi_k(t) \, dt.$$

It is possible to give a construction of the Fourier transform on $L^2(-\infty, \infty)$ different from that presented in Chapter 5 on the basis of our considerations. Namely, we represent every function $f \in L^2$ by the series (4), and define its Fourier transform by the series

$$\hat{f}(t) \overset{L^2}{=} \sum_{k=0}^{\infty} (-i)^k \frac{c_k}{\alpha_k^2} \varphi_k(t).$$

The proof that this definition is equivalent to the Plancherel one is left to the reader.

**4.** PROBLEMS. 1. Prove that the Hermite functions $\varphi_k(t)$ are eigenfunctions of the differential operator $L = -d^2/dt^2 + t^2$, namely, $L\varphi_k = (2k+1)\varphi_k$. The collection of numbers $\{2k+1\}_{k=0}^\infty$ is the full spectrum of $L$.

2. Prove that the polynomials $H_k(t)$ satisfy the recursion relation

$$H_k(t) + 2(k-1)H_{k-2}(t) = 2tH_{k-1}(t).$$

CHAPTER 7

# Spherical Functions

**1.** Let $x \in E_n$ be an arbitrary vector, $x = r\xi$, where $r \geq 0$ and $\xi \in S_{n-1}$. Instead of coordinates $\xi_k$ $(k = 1, \dots, n)$ connected by the relation $\sum_1^n \xi_k^2 = 1$ one frequently introduces the angles $\nu_1, \dots, \nu_{n-1}$ as independent coordinates, setting

$$\xi_1 = \sin \nu_1 \sin \nu_2 \cdots \sin \nu_{n-2} \cdot \sin \nu_{n-1},$$
$$\xi_j = \sin \nu_1 \cdot \sin \nu_2 \cdots \sin \nu_{n-j} \cdot \cos \nu_{n-j+1} \qquad (j = 2, 3, \dots, n-1),$$
$$\xi_n = \cos \nu_1.$$

To obtain the whole of the sphere the angle $\nu_{n-1}$ here must run through the interval $[0, 2\pi)$, and the remaining angles through the interval $[0, \pi]$. In the special case $n = 3$ the formulas have the form

$$\xi_1 = \sin \nu_1 \cdot \sin \nu_2, \quad \xi_2 = \sin \nu_1 \cdot \cos \nu_2, \quad \xi_3 = \cos \nu_1,$$

and here $\nu_2$ is the longitude $(0 \leq \nu_2 < 2\pi)$ and $\nu_1$ the latitude $(0 \leq \nu_1 \leq \pi)$. For $n > 2$ the role of the longitude is played by $\nu_{n-1}$, and the role of the latitudes by the system $\{\nu_1, \dots, \nu_{n-2}\}$. It is well known that the Laplace operator can be represented in the form

$$\Delta = \frac{1}{r^2} \frac{\partial}{\partial r} \left( r^2 \frac{\partial}{\partial r} \right) + \frac{1}{r^2} \left\{ \frac{1}{\sin \nu_1} \cdot \frac{\partial}{\partial \nu_1} \left( \sin \nu_1 \frac{\partial}{\partial \nu_1} \right) + \frac{1}{\sin^2 \nu_1} \frac{\partial^2}{\partial \nu_2^2} \right\}$$

in three-dimensional space. In the case of an arbitrary $n \geq 2$ the Laplace operator can be represented in the form

$$\Delta = \frac{1}{r^{n-1}} \frac{\partial}{\partial r} + \left( r^{n-1} \frac{\partial}{\partial r} \right) + \frac{1}{r^2} \tilde{\Delta} \tag{1}$$

in $E_n$, where $\tilde{\Delta}$ is a second-order linear differential operator with respect to the variables $\nu_k$ $(k = 1, \dots, n-1)$ or with respect to the variables $\xi_k$ $(k = 1, \dots, n)$, i.e., with respect to the coordinates on $S_{n-1}$. The operator $\tilde{\Delta}$ is usually called the *Beltrami operator.* It follows directly from the representation (1) that for $n > 2$ the only harmonic functions depending only on $r$ are $C$ and $C \cdot r^{2-n}$ ($C$ a constant). Instead of $r$ it is possible to take the distance of the variable point $x$ from an arbitrarily chosen fixed point $a$. Then we get the nontrivial solution $1/|x-a|^{n-2}$ $(x \neq a)$. In view of its role in the theory of harmonic functions, this solution is called a *fundamental solution* of the Laplace equation in $E_n$. For $n = 3$ a fundamental solution is a Newton potential.

For $n = 2$ it follows from (1) that the only solutions depending only on $r$ and $C$ and $C \cdot \ln(1/r)$. The logarithmic potential is the fundamental solution.

DEFINITION. A *kth-degree ball function* $(k = 0, 1, \dots)$ on $E_n$ is defined to be any $k$th degree homogeneous polynomial satisfying the Laplace equation in $E_n$. The restriction of a $k$th-degree ball function to $S_{n-1}$ is called a *kth-degree spherical function* in $E_n$.

If $\Phi_k(x)$ is a $k$th-degree ball function, then $\Phi_k(x) = r^k \Phi_k(\xi)$, where $\Phi_k(\xi)$ is a $k$th-degree spherical function. We indicate the dependence of functions on the dimension $n$ of the space only when necessary.

Substituting $\Phi_k(x)$ into the equation $\Delta u = 0$, where the operator $\Delta$ is taken in the form (1), we find easily that the spherical functions on $E_n$ satisfy

$$\tilde{\Delta}\Phi_k(\xi) + k(n+k-2)\Phi_k(\xi) = 0, \tag{2}$$

i.e., they are eigenfunctions of the Beltrami operator.

**2.** We find the Fourier transform of the function $e^{-|x|^2/2}\Phi_k(x)$, where $\Phi_k(x)$ is a $k$th-degree ball function on $E_n$. For this we write the equality (see Chapter 2, §3)

$$\frac{1}{(\sqrt{2\pi})^n} \int_{E_n} e^{-|y|^2/2} e^{-i(t,y)}\, dy = e^{-|t|^2/2}.$$

From this it is possible to get the new equality

$$\frac{1}{(\sqrt{2\pi})^n} \int_{E_n} e^{-|y|^2/2} e^{-i(t,y)} \Phi_k(y)\, dy = e^{-|t|^2/2} Q_k(t)$$

by differentiating with respect to $t_1, \dots, t_n$ and taking linear combinations (with complex coefficients), where $\Phi_k(y)$ is a previously given $k$th-degree ball function, and $Q_k(t)$ is obviously some $k$th-degree polynomial which we want to find. With this goal we rewrite our relation in the form

$$Q_k(t) = \frac{1}{(\sqrt{2\pi})^n} \int_{E_n} e^{-1/2 \sum_{k=1}^n (y_k + it_k)^2} \Phi_k(y)\, dy.$$

Both sides are analytic functions of the parameters $t_1, \dots, t_n$, and the equality holds also when we give them purely imaginary values. Setting $t_j = ix_j$

$(j = 1, \dots, n)$, we represent our relation in the form

$$Q_k(ix) = \frac{1}{(\sqrt{2\pi})^n} \int_{E_n} e^{-1/2|y-x|^2} \Phi_k(y)\, dy.$$

Using the mean value theorem for harmonic functions, we find that

$$\begin{aligned} Q_k(ix) &= \frac{1}{(\sqrt{2\pi})^n} \int_0^\infty e^{-r^2/2}\, dr \int_{S_{n-1}^x(r)} \Phi_k(y)\, d\sigma \\ &= \frac{1}{(\sqrt{2\pi})^n} \int_0^\infty e^{-r^2/2} \Phi_k(x) \cdot |S_{n-1}(r)|\, dr \\ &= \Phi_k(x) \frac{1}{(\sqrt{2\pi})^n} \int_{E_n} e^{-|y|^2/2}\, dy = \Phi_k(x). \end{aligned}$$

Thus, $Q_k(ix) = \Phi_k(x)$, and this relation between the two analytic functions is preserved for purely imaginary values $x = -it$. Using the fact that $\Phi_k$ is homogeneous of degree $k$, we get that $Q_k(t) = (-i)^k \Phi_k(t)$. It now follows from the definition of $Q_k(t)$ that

$$\frac{1}{(\sqrt{2\pi})^n} \int_{E_n} e^{-|y|^2/2} e^{-i(t,y)} \Phi_k(y)\, dy = (-i)^k \Phi_k(t) e^{-|t|^2/2}.$$

This result is called the *Hecke identity.*

**3.** Spherical functions in $E_n$ play the role of periodic functions (i.e., of point functions on $S_1$) in the consideration of point functions on $S_{n-1}$. In this chapter we thus want to study certain questions connected with series expansions in spherical functions. The Poisson integral plays an important role in our considerations. We first take the function

$$\frac{1 - w^2}{(1 - 2wz + w^2)^\lambda}$$

of the complex variable $w$ and of the real parameters $z$ and $\lambda > 0$. This function admits an expansion in a Maclaurin series in the disk $|w| < 1$:

$$\frac{1 - w^2}{(1 - 2wz + w^2)^\lambda} = \sum_0^\infty \alpha_k(z) w^k; \qquad \alpha_0(z) = 1. \tag{3}$$

It is clear from the binomial series expansion

$$\begin{aligned} \frac{1 - w^2}{(1 - 2wz + w^2)^\lambda} &= \frac{1 - w^2}{(1 - [2wz - w^2])^\lambda} \\ &= (1 - w^2) \sum_{j=0}^\infty \binom{-\lambda}{j} (-1)^j (2wz - w^2)^j \end{aligned}$$

that the coefficient $\alpha_k(z)$ in (3) is a polynomial precisely of degree $k$.([18]) Here we do not indicate the dependence of the coefficient $\alpha_k(z)$ on $\lambda$. Since replacement of $w$ by $-w$ and of $z$ by $-z$ leaves the left-hand side of (3) unchanged, it follows that $\alpha_k(z) = (-1)^k \alpha_k(-z)$, i.e., $\alpha_k(z)$ is an even polynomial for even $k$ and an odd polynomial for odd $k$. Turning now to the Poisson kernel $H(x, \eta)$ in $E_n$, we can write the expansion (where $z = \cos\gamma = (\xi, \eta)$ and $r < 1$)

$$\frac{1 - r^2}{(1 - 2zr + r^2)^{n/2}} = \sum_{k=0}^{\infty} \alpha_k^{(n)}(\cos\gamma) r^k; \tag{4}$$

the dependence of the coefficients on the dimension of the space is now indicated. The coefficient of $r^k$ in this expansion is the polynomial

$$\alpha_k^{(n)}(\cos\gamma) = A \cdot (\xi, \eta)^k + A' \cdot (\xi, \eta)^{k-2} + \cdots.$$

The function

$$h_k^{(n)}(x, \eta) = A \cdot (x, \eta)^k + A' \cdot (x, \eta)^{k-2} r^2 + \cdots$$

is a homogeneous polynomial of degree $k$ in $x$. Further,

$$h_k^{(n)}(r\xi, \eta) = r^k \alpha_k^{(n)}(\cos\gamma),$$

so that the series (4) can be written in the form

$$\frac{1 - |x|^2}{[1 - 2(x, \eta) + |x|^2]^{n/2}} = \sum_{k=0}^{\infty} h_k^{(n)}(x, \eta). \tag{$4_1$}$$

The left-hand side of $(4_1)$ is clearly analytic in a neighborhood of $x = 0$, and the expansion $(4_1)$ is obtained from its Taylor expansion by combining monomials of the same degree. Therefore, the series can be differentiated termwise any number of times with preservation of absolute and uniform convergence in a neighborhood of the origin. Since the left-hand side of $(4_1)$ is a harmonic function,

$$0 = \sum_{k=0}^{\infty} \Delta_x h_k^{(n)}(x, \eta). \tag{$4_2$}$$

Since the $k$th term on the right-hand side is a homogeneous polynomial in $x_1, \ldots, x_n$ of degree $k - 2$, each term of the right-hand side of $(4_2)$ is in fact equal to 0. Hence, $h_k^{(n)}(x, \eta)$ is a ball function of degree $k$ in $x$, and $h_k^{(n)}(\xi, \eta) = \alpha_k^{(n)}(\cos\gamma)$ is a spherical function of degree $k$ both in $\xi$ and in $\eta$. This function is called a *zonal spherical function of degree* $k$ in $E_n$.([19])

([18])With leading coefficient $2^k \lambda(\lambda + 1) \cdots (\lambda + k - 1)/k!$ $(k \geq 1)$.

([19])By analogy with the terminology for the case $n = 3$.

**4.** There are only finitely many linearly independent real $k$th-degree spherical functions in $E_n$ for each $k$; denote this number by $N_k^{(n)}$. It is easy to see that $N_k^{(1)} = 2$. Indeed, the only homogeneous analytic function of degree $k$ on the plane is $(\alpha + i\beta)z^k$. Therefore, for any $k$ we can take $(z^k + \overline{z}^k)/2$ and $(z^k - \overline{z}^k)/2i$ as the real linearly independent spherical functions on the plane. Below, we get an expression for $N_k^{(k)}$ in terms of $n$ and $k$. For any $n$ we introduce a basis of real spherical functions of degree $k$ on $E_n$ that is orthonormal with respect to the inner product

$$\langle f, g\rangle = \frac{1}{|S_{n-1}|}\int_{S_{n-1}} f(\eta)g(\eta)\,d\sigma.$$

Denote this basis by $\{\Phi_{kj}^{(n)}(\xi)\}_{j=1}^{N_k^{(n)}}$ $(k = 0, 1, 2, \dots)$. We can now represent a zonal function of degree $k$ in the form

$$\alpha_k^{(n)}(\cos\gamma) = \sum_{j,l=1}^{N_k^{(n)}} c_{jl}\Phi_{kj}^{(n)}(\xi)\Phi_{kl}^{(n)}(\eta), \tag{5}$$

where $c_{jl} = c_{lj}$ by symmetry.([20])

We write the ball function $r^k\Phi_{kj}^{(n)}(\xi)$ as a Poisson integral

$$\begin{aligned} r^k\Phi_{kj}^{(n)}(\xi) &= \frac{1}{|S_{n-1}|}\cdot\int_{S_{n-1}} \Phi_{kj}^{(n)}(\eta)H(r\xi,\eta)\,d\sigma_\eta \\ &= \frac{1}{|S_{n-1}|}\cdot\int_{S_{n-1}} \Phi_{kj}^{(n)}(\eta)\sum_{l=0}^{\infty} r^l\alpha_l^{(n)}(\cos\gamma)\,d\sigma_\eta. \end{aligned}$$

This gives us the following relations:

$$\int_{S_{n-1}} \Phi_{kj}^{(n)}(\eta)\alpha_l^{(n)}(\cos\gamma)\,d\sigma_\eta = 0 \qquad (l \neq k),$$

$$\frac{1}{|S_{n-1}|}\cdot\int_{S_{n-1}} \Phi_{kj}^{(n)}(\eta)\alpha_k^{(n)}(\cos\gamma)\,d\sigma_\eta = \Phi_{kj}^{(n)}(\xi).$$

By (5), the second of these can be represented in the form

$$\frac{1}{|S_{n-1}|}\int_{S_{n-1}} \Phi_{kj}^{(n)}(\eta)\sum_{i,l=1}^{N_k^{(n)}} c_{il}\Phi_{ki}^{(n)}(\xi)\Phi_{kl}^{(n)}(\eta)\,d\sigma_\eta = \Phi_{kj}^{(n)}(\xi),$$

which implies that $c_{ij} = \delta_{ij}$. Thus,

$$\alpha_k^{(n)}(\cos\gamma) = \sum_{j=1}^{N_k^{(n)}} \Phi_{kj}(\xi)\Phi_{kj}(\eta).$$

([20])For the time being we must regard $c_{jl}$ as a function of $n$ and $k$. However, we shall soon see that $c_{jl} = \delta_{jl}$.

If we set $\xi = \eta$ here, we get

$$\alpha_k^{(n)}(1) = \sum_{j=1}^{N_k^{(n)}} [\Phi_{kj}^{(n)}(\eta)]^2.$$

Integrating both sides over $S_{n-1}$, we get the formula $N_k^{(n)} = \alpha_k^{(n)}(1)$ for the dimension of the space of real $k$th-degree spherical functions in $E_n$. Formula (3) with $\lambda = n/2$ gives us

$$\frac{1+w}{(1-w)^{n-1}} = \sum_{k=0}^{\infty} N_k^{(n)} w^k,$$

which implies that

$$N_k^{(2)} = \begin{cases} 1 & (k=0), \\ 2 & (k>0), \end{cases} \qquad N_k^{(n)} = \frac{(n+k-3)!(n+2k-2)}{k!(n-2)!} \qquad (n \geq 3; k \geq 0).$$

**5.** We now turn to a study of the polynomials $\alpha_k^{(n)}(t)$ in the interval $(-1, 1)$, and prove that they are orthogonal with respect to a certain weight. As a starting point we take the equalities

$$\frac{1}{|S_{n-1}|} \int_{S_{n-1}} \alpha_k^{(n)}(\cos\gamma) \cdot \alpha_l^{(n)}(\cos\gamma)\, d\sigma_\eta = \alpha_k^{(n)}(1) \cdot \delta_{kl}, \tag{6}$$

which follow from the results in §4. Since the integrals in (6) do not depend on $\xi$, it can be assumed that $\xi = \{0, \dots, 0, 1\}$, and then $\cos\gamma = \eta_n$. The intersection of the sphere $\eta_1^2 + \cdots + \eta_n^2 = 1$ with the plane $\eta_n = t$ is an $(n-2)$-dimensional sphere of radius $\sqrt{1-t^2}$. For any function $f(t)$, $t \in (-1, 1)$,

$$\begin{aligned} \int_{S_{n-1}} f(\cos\gamma)\, d\sigma_\eta &= \int_0^\pi f(\cos\gamma)|S_{n-2}(\sin\gamma)|\, d\gamma \\ &= |S_{n-2}| \int_0^\pi f(\cos\gamma)(\sin\gamma)^{n-2}\, d\gamma \\ &= |S_{n-2}| \int_{-1}^1 f(t)(\sqrt{1-t^2})^{n-3}\, dt. \end{aligned}$$

Recalling that $|S_{n-1}| = 2(\sqrt{\pi})^n/\Gamma(n/2)$, we can rewrite (6) in the form

$$\frac{\Gamma(n/2)}{\Gamma(1/2)\Gamma((n-1)/2)} \cdot \int_{-1}^1 \alpha_k^{(n)}(t)\alpha_l^{(n)}(t)(1-t^2)^{(n-3)/2}\, dt = \delta_{kl} \cdot \alpha_k^{(n)}(1).$$

As we see, the polynomials $\alpha_k^{(n)}(t)$ are orthogonal with respect to the weight $(1-t^2)^{(n-3)/2}$. They are a special case of the Jacobi orthogonal polynomials, and are called the *Gegenbauer polynomials.*

**6.** Assume now that $\varphi(\xi)$ is a given real and (for simplicity) continuous point function on $S_{n-1}$. We can associate with it the series of spherical functions

$$\varphi(\xi) \sim \sum_{k=0}^{\infty} \sum_{j=1}^{N_k^{(n)}} a_{kj} \Phi_{kj}^{(n)}(\xi), \tag{7}$$

where

$$a_{kj} = \frac{1}{|S_{n-1}|} \int_{S_{n-1}} \varphi(\eta) \Phi_{kj}^{(n)}(\eta)\, d\sigma. \tag{8}$$

Let us prove the equality

$$\frac{1}{|S_{n-1}|} \int_{S_{n-1}} |\varphi(\xi)|^2 \, d\sigma = \sum_{k=0}^{\infty} \sum_{j=1}^{N_k^{(n)}} |a_{kj}|^2, \tag{9}$$

which we call the *Parseval equality*. For this we take the Poisson integral

$$u(r\xi) = \frac{1}{|S_{n-1}|} \cdot \int_{S_{n-1}} \frac{1 - r^2}{[1 - 2r\cos\gamma + r^2]^{n/2}} \varphi(\eta)\, d\sigma_\eta.$$

Expanding the Poisson kernel in a series of spherical functions for $0 < r < 1$, we get from (8) and the results in §4 that

$$u(r\xi) = \sum_{k=0}^{\infty} r^k \sum_{j=1}^{N_k^{(n)}} a_{kj} \Phi_{kj}^{(n)}(\xi).$$

Multiplying both sides by $\varphi(\xi)$ and integrating over $S_{n-1}$, we find from this that

$$\frac{1}{|S_{n-1}|} \int_{S_{n-1}} u(r\xi)\varphi(\xi)\, d\sigma = \sum_{k=0}^{\infty} r^k \sum_{j=1}^{N_k^{(n)}} |a_{kj}|^2.$$

Taking the limit as $r \nearrow 1$, we get the Parseval equality (9). It is valid for functions $\varphi \in L^2(S_{n-1})$.

We make an important remark about the spherical function series in (7). The terms in it are grouped; the sum of the terms in the $k$th group is

$$\sum_{j=1}^{N_k^{(n)}} a_{kj} \Phi_{kj}^{(n)}(\xi) = \Phi_k^{(n)}(\xi).$$

The function $\Phi_k^{(n)}(\xi)$ is the *kth spherical harmonic* appearing in the expansion. It is clear that the individual terms of which it is made up may not be known. In this connection it is worth mentioning that, by (8),

$$\Phi_k^{(n)}(\xi) = \frac{1}{|S_{n-1}|} \int_{S_{n-1}} \varphi(\eta) \sum_{j=1}^{N_k^{(n)}} \Phi_{kj}^{(n)}(\xi) \Phi_{kj}^{(n)}(\eta)\, d\sigma_\eta$$
$$= \frac{1}{|S_{n-1}|} \int_{S_{n-1}} \varphi(\eta) \alpha_k^{(n)}(\cos\gamma)\, d\sigma_\eta,$$

i.e., every harmonic is represented as an integral in terms of Gegenbauer polynomials.

7. Problems. 1. Prove that the Laplacian in $E_n$ can be represented in the form

$$\frac{1}{r^2}\left[\left(r\frac{\partial}{\partial r}\right)^2 + (n-2)\left(r\frac{\partial}{\partial r}\right) + \tilde{\Delta}\right],$$

and derive from this the following proposition: *if $u$ is a harmonic function, then so is $v = r(\partial u/\partial r)$.*

2. Using the proposition in the preceding problem, prove that the Poisson kernel

$$\frac{1-r^2}{[1-2r(\xi,\eta)+r^2]^{n/2}}$$

is harmonic.

3. Let $u(x)$ be harmonic everywhere in $E_n$, and suppose that for each $x \in E_n$

$$u(x) \le C(1+|x|^p),$$

where $C$ and $p$ are positive constants. Prove that under this condition $u(x)$ is a harmonic polynomial of degree $\le p$ (a generalization of the Picard-Liouville theorem).

*Solution.* Write the expansion (valid for $R > 0$ and any $\xi \in S_{n-1}$)

$$u(R\xi) = \sum_{k=0}^{\infty} R^k \Phi_k^{(n)}(\xi),$$

where, by the remark at the end of §6, the spherical function $\Phi_k^{(n)}(\xi)$ can be represented in terms of a zonal spherical function in the form

$$R^k \Phi_k^{(n)}(\xi) = \frac{1}{|S_{n-1}|} \int_{S_{n-1}} u(R\eta) \alpha_k^{(n)}(\cos\gamma)\, d\sigma_\eta.$$

This implies the inequality

$$R^k \cdot [\Phi_k^{(n)}(\xi)| \le \frac{M_k^{(n)}}{|S_{n-1}|} \int_{S_{n-1}} |u(R\eta)|\, d\sigma,$$

where

$$M_k^{(n)} = \max_{-1 \le t \le 1} |\alpha_k^{(n)}(t)|,$$

or the inequality

$$R^k \cdot |\Phi_k^{(n)}(\xi)| \le \frac{M_k^{(n)}}{|S_{n-1}} \int_{S_{n-1}} \{|u(R\eta)| + u(R\eta)\}\, d\sigma - M_k^{(n)} \cdot u(0).$$

But, on the other hand,

$$|u(R\eta)| + u(R\eta) = 2\max\{u(R\eta), 0\}.$$

Therefore, by a condition of the problem,

$$R^k \cdot |\Phi_k^{(n)}(\xi)| \le 2 \cdot M_k^{(n)} \cdot C \cdot (1+R)^p - M_k^{(n)} \cdot u(0),$$

from which it follows that $\Phi_k^{(n)}(\xi) = 0$ for $k > p$.

4. Let $f(t)$ $(-1 \le t \le 1)$ be a given continuous function. Prove that the equality

$$\frac{1}{|S_{n-1}|} \int_{S_{n-1}} f(\cos\gamma) \Phi_k^{(n)}(\eta)\, d\sigma_\eta = \lambda_k \Phi_k^{(n)}(\xi)$$

holds for any spherical function $\Phi_k^{(n)}(\xi)$ of degree $k$ in $E_n$; this shows that the spherical functions of degree $k$ $(k = 0, 1, \dots)$ are eigenfunctions of the kernel $f(\cos\gamma)$. Prove also that the eigenvalues $\lambda_k$ are determined by the formulas

$$\lambda_k = \frac{\Gamma(n/2)}{\Gamma(1/2) \cdot \Gamma((n-1)/2)} \int_{-1}^{1} f(t) \frac{\alpha_k^{(n)}(t)}{\alpha_k^{(n)}(1)} (1-t^2)^{(n-3)/2}\, dt,$$

i.e., they are the coefficients in the expansion

$$f(t) \sim \sum_{k=0}^{\infty} \lambda_k \alpha_k^{(n)}(t).$$

5. Apply the general result in the preceding problem to the function $f(t) = e^{-irt}$ $(-1 \le t \le 1)$, where the parameter $r$ is positive. Thus, now $\lambda_k = \lambda_k(r)$ in the formula

$$\frac{1}{|S_{n-1}|} \int_{S_{n-1}} e^{-ir\cos\gamma} \Phi_k^{(n)}(\eta)\, d\sigma_\eta = \lambda_k \Phi_k^{(n)}(\xi).$$

Using the fact that the function $e^{-i(x,\eta)}$ satisfies the equation $\Delta e^{-i(x,\eta)} = -e^{-i(x,\eta)}$, for each fixed $\eta$ $(|\eta| = 1)$, prove the equality

$$\lambda_k(r) = (-i)^k 2^{n/2} \Gamma\left(\frac{n}{2}\right) \frac{J_{k+n/2-1}(r)}{r^{n/2-1}};$$

in particular, for $n = 3$ obtain the equality

$$\frac{1}{\sqrt{2\pi}} \int_{-1}^{1} P_k(t) e^{-irt}\, dt = (-i)^k \frac{J_{k+1/2}(r)}{\sqrt{r}},$$

where $P_k$ is the Legendre polynomial, and $J_\nu$ the Bessel function.

6. Prove that the Legendre polynomial $P_k(t)$ satisfies the differential equation

$$-\frac{d}{dt}(1-t^2)\frac{dy}{dt} = k(k+1)y.$$

7. Prove that the Legendre polynomials satisfy the recursion relation

$$(k+1)P_{k+1}(t) + kP_{k-1}(t) = (2k+1)tP_k(t).$$

CHAPTER 8

# Positive Definite Functions

**1.** DEFINITION. A complex-valued function $f(x)$ $(x \in E_n)$ is said to be *positive definite* if the inequality

$$\sum_{\alpha,\beta=1}^{N} f(x^\alpha - x^\beta)\xi_\alpha\overline{\xi}_\beta \geq 0 \tag{1}$$

holds for any collection of points $x^k \in E_n$ $(k = 1, 2, \ldots, N < \infty)$ and any complex numbers $\xi_k$. Setting $N = 1$ in (1), we find that $f(0) \cdot |\xi|^2 \geq 0$, which implies that $f(0) \geq 0$ for every positive definite function. Next, let $N = 2$. The inequality (1) takes the form

$$f(0)\xi_1\overline{\xi}_1 + f(x^1 - x^2)\xi_1\overline{\xi}_2 + f(x^2 - x^1)\xi_2\overline{\xi}_1 + f(0)\xi_2\overline{\xi}_2 \geq 0.$$

By replacing $x^1 - x^2$ by $x$, we find from this that $f(-x) = \overline{f(x)}$. Moreover, the discriminant

$$\begin{vmatrix} f(0) & f(x) \\ \overline{f(x)} & f(0) \end{vmatrix}$$

must be nonnegative; we get the property $|f(x)| \leq f(0)$, and hence every positive definite function is bounded. Finally, let $N = 3$ and take $x^3 = 0$, $\xi_1 = -\xi_2 = \xi$, and $\xi_3 = \eta$. Then inequality (1) takes the form

$$\begin{aligned} &f(0) \cdot |\xi|^2 - f(x^1 - x^2) \cdot |\xi|^2 + f(x^1) \cdot \xi\overline{\eta} - f(x^2 - x^1) \cdot |\xi|^2 \\ &\quad + f(0) \cdot |\xi|^2 - f(x^2) \cdot \xi\overline{\eta} + f(-x^1) \cdot \eta\overline{\xi} - f(-x^2)\eta\overline{\xi} + f(0) \cdot |\eta|^2 \geq 0, \end{aligned}$$

or

$$\begin{aligned} &[f(0)] \cdot |\eta|^2 + 2 \cdot \left[f(0) - \frac{f(x^1 - x^2) + f(x^2 - x^1)}{2}\right] \cdot |\xi^2| \\ &\quad + [f(x^1) - f(x^2)] \cdot \xi\overline{\eta} + [\overline{f(x^1) - f(x^2)}] \cdot \overline{\xi}\eta \geq 0, \end{aligned}$$

which implies that

$$|f(x^1) - f(x^2)|^2 \leq 2f(0)\left[f(0) - \frac{f(x^1 - x^2) + f(x^2 - x^1)}{2}\right].$$

Since

$$f(0) - \frac{f(x^1 - x^2) + f(x^2 - x^1)}{2} = \operatorname{Re}[f(0) - f(x^1 - x^2)],$$

the continuity of the real part of a positive definite function at 0 gives us the uniform continuity of the function itself in the whole space. In what follows we consider only continuous positive definite functions. Note that for these functions there is a continuous analogue of (1), namely

$$\int_{E_n}\int_{E_n} f(x-y)\rho(x)\overline{\rho(y)}\,dx\,dy \geq 0, \tag{1'}$$

where $\rho(x)$ is an arbitrary continuous function in $L^1(E_n)$. Let us verify this assertion for $n = 1$. In this case

$$\begin{aligned}
&\int_{-\infty}^{\infty}\int_{-\infty}^{\infty} f(x-y)\rho(x)\overline{\rho(y)}\,dx\,dy \\
&= \lim_{A\to\infty}\int_{-A}^{A}\int_{-A}^{A} f(x-y)\rho(x)\overline{\rho(y)}\,dx\,dy \\
&= \lim_{A\to\infty}\lim_{N\to\infty}\sum_{\alpha,\beta=-N}^{N} f\left(\frac{\alpha A - \beta A}{N}\right)\rho\left(\frac{\alpha A}{N}\right)\overline{\rho\left(\frac{\beta A}{N}\right)}\frac{A^2}{N^2},
\end{aligned}$$

and the double sum on the right-hand side is nonnegative by (1).

**2.** The main theorem in the theory of positive definite functions is a theorem giving an integral representation of these functions. For $n = 1$ this theorem was proved almost simultaneously by Khintchine and Bochner. It was proved a little later by Bochner for arbitrary $n > 1$.

Let $f(x)$ $(x \in E_n)$ be a given continuous positive definite function. We write inequality (1′) with

$$\rho(x) = e^{-2\varepsilon|x|^2}e^{-i(t,x)},$$

and get that

$$\int_{E_n}\int_{E_n} f(x-y)e^{-2\varepsilon|x|^2}e^{-2\varepsilon|y|^2}e^{-i(t,x)}e^{i(t,y)}\,dx\,dy \geq 0.$$

Let us make the change of variables $x - y = u$, $x + y = v$. Computation of the Jacobian gives us that $dx\,dy$ must be replaced by $2^{-n}\,du\,dv$. Further, $|u|^2 + |v|^2 = 2|x|^2 + 2|y|^2$. Therefore, our inequality takes the form

$$\frac{1}{2^n}\int_{E_n}\int_{E_n} f(u)e^{-\varepsilon|u|^2}e^{-\varepsilon|v|^2}e^{-i(t,u)}\,du\,dv \geq 0,$$

or

$$\int_{E_n} e^{-\varepsilon|v|^2}\,dv \int_{E_n} f(u)e^{-\varepsilon|u|^2}e^{-i(t,u)}\,du \geq 0,$$

from which, adding a constant factor, we get

$$\frac{1}{(\sqrt{2\pi})^n}\int_{E_n} f(u)e^{-\varepsilon|u|^2}e^{-i(t,u)}\,du \geq 0.$$

If we introduce the bounded absolutely integrable function (with parameter $\varepsilon > 0$) $f_\varepsilon(u) = f(u)e^{-\varepsilon|u|^2}$, then the last inequality means that the Fourier transform $\hat{f}_\varepsilon(t)$ of this function is everywhere nonnegative. By Theorem 2 in Chapter 3, $\hat{f}_\varepsilon \in L^1$, and we have the inversion formula

$$f(u)e^{-\varepsilon|u|^2} = \frac{1}{(\sqrt{2\pi})^n}\int_{E_n} e^{i(u,t)}\hat{f}_\varepsilon(t)\,dt. \tag{2}$$

The left-hand side has a limit as $\varepsilon \searrow 0$, but it is not clear how to realize the passage to the limit on the right-hand side of (2).

Let us consider in detail the case when $n = 1$ and, hence,

$$f(u)e^{-\varepsilon u^2} = \frac{1}{\sqrt{2\pi}}\int_{-\infty}^{\infty} e^{iut}\hat{f}_\varepsilon(t)\,dt \qquad (-\infty < u < \infty). \tag{3}$$

We introduce the function

$$\omega_\varepsilon(t) = \frac{1}{\sqrt{2\pi}}\int_{-\infty}^{t} \hat{f}_\varepsilon(v)\,dv. \tag{4}$$

It follows from the positivity of $\hat{f}_\varepsilon(t)$ and (3) that for any $\varepsilon > 0$ the function $\omega_\varepsilon(t)$ is monotone and $0 \leq \omega_\varepsilon(t) \leq f(0)$. Formula (3) can be written in the form

$$f(u)e^{-\varepsilon u^2} = \int_{-\infty}^{\infty} e^{iut}\,d\omega_\varepsilon(t). \tag{5}$$

We now use Helly's selection theorem([21]) which implies that for the family $\{\omega_\varepsilon(t)\}$ with parameter $\varepsilon > 0$ there exist a monotone function $\omega(t)$, $0 \leq \omega(t) \leq f(0)$, and a sequence $\varepsilon_i \to 0$ $(i \to \infty)$ such that $\lim_{i\to\infty}\omega_{\varepsilon_i}(t) = \omega(t)$ for all $t$. We next show that for $\varepsilon = \varepsilon_i \to 0$ the equality $f(u) = \int_{-\infty}^{\infty} e^{iut}\,d\omega(t)$ follows from (5). To do this we first need Helly's limit theorem,([21]) which implies that

$$\lim_{i\to\infty}\int_{-A}^{B}\varphi(t)\,d\omega_{\varepsilon_i}(t) = \int_{-A}^{B}\varphi(t)\,d\omega(t),$$

([21])See, for example, A. N. Kolmogorov and S. V. Fomin, *Elements of the theory of functions and of functional analysis*, 3rd ed., "Nauka", Moscow, 1972, Chapter VI, §6 (English transl. of 1st ed., Vol. II, Greylock Press, Albany, N.Y., 1961); or I. P. Natanson, *Theory of functions of a real variable*, 2nd rev. ed., GITTL, Moscow, 1957, Chapter VIII (English transl. of 1st ed., Vol. 1, Ungar, New York, 1955).

where $\varphi(t)$ is any continuous function on the finite interval $[-A, B]$. Second, there is an extra difficulty, because in our case the interval of integration is infinite. In this connection we rewrite (5) in the form

$$f(u)e^{-\varepsilon u^2} - \int_{-A}^{B} e^{iut}\, d\omega_\varepsilon(t) = \int_{-\infty}^{-A} e^{iut}\, d\omega_\varepsilon(t) + \int_{B}^{\infty} e^{iut}\, d\omega_\varepsilon(t),$$

where $A,\ B > 0$. For the right-hand side of this formula we take the majorant

$$\left| \int_{-\infty}^{-A} e^{iut}\, d\omega_\varepsilon(t) + \int_{B}^{\infty} e^{iut}\, d\omega_\varepsilon(t) \right| \le \int_{|t|>C} d\omega_\varepsilon(t),$$

where $C = \min\{A, B\}$.

LEMMA. *For any $\delta > 0$ the inequality*

$$\int_{|t|>C} d\omega_\varepsilon(t) < \delta$$

*holds for all $\varepsilon > 0$ if $C \ge C_\delta$.*

Assume that this lemma has been proved. Taking an arbitrary $\delta > 0$, we find a $C_\delta$ corresponding to it. Then for all $\varepsilon > 0$ the inequality

$$\left| f(u)e^{-\varepsilon u^2} - \int_{-A}^{B} e^{iut}\, d\omega_\varepsilon(t) \right| \le \delta$$

holds under the condition $\min\{A, B\} \ge C_\delta$, and the right-hand side does not depend on $\varepsilon$. Take $\varepsilon = \varepsilon_i$ and let $i$ go to $\infty$. By Helly's limit theorem,

$$\left| f(u) - \int_{-A}^{B} e^{iut}\, d\omega(t) \right| \le \delta.$$

For an arbitrary $\delta > 0$ this inequality holds if

$$\min\{A, B\} \ge C_\delta.$$

But this means that

$$f(u) = \int_{-\infty}^{\infty} e^{iut}\, d\omega(t).$$

This equality is the Bochner-Khintchine representation of a continuous positive definite function.

PROOF OF THE LEMMA. We derive two relations from (5). First, we write the equality

$$f(0) = \int_{-\infty}^{\infty} d\omega_\varepsilon(t),$$

and, second, we multiply both sides of (5) by $(1/2\gamma)du$ and integrate with respect to $u$ from $-\gamma$ to $\gamma$ $(\gamma > 0)$, after which we have

$$\frac{1}{2\gamma}\int_{-\gamma}^{\gamma} f(u)e^{-\varepsilon u^2}\,du = \int_{-\infty}^{\infty} \frac{\sin\gamma t}{\gamma t}\,d\omega_\varepsilon(t).$$

It follows that

$$\int_{-\infty}^{\infty}\left[1-\frac{\sin\gamma t}{\gamma t}\right]d\omega_\varepsilon(t) = f(0) - \frac{1}{2\gamma}\int_{-\gamma}^{\gamma} f(u)e^{-\varepsilon u^2}\,du. \qquad (*)$$

The expression $1 - \sin\gamma t/\gamma t$ is always nonnegative, and if $\gamma|t| > 2$, then it is greater than $1/2$. Therefore, the left-hand side of $(*)$ is greater than $\frac{1}{2}\int_{|t|>2/\gamma} d\omega_\varepsilon(t)$. Hence,

$$\begin{aligned}\frac{1}{2}\int_{|t|<2/\gamma} d\omega_\varepsilon(t) &< f(0) - \frac{1}{2\gamma}\int_{-\gamma}^{\gamma} f(u)e^{-\varepsilon u^2}\,du \\ &= \left\{f(0) - \frac{1}{2\gamma}\int_{-\gamma}^{\gamma} f(u)\,du\right\} + \frac{1}{2\gamma}\int_{-\gamma}^{\gamma}(1-e^{-\varepsilon u^2})f(u)\,du = I_1 + I_2.\end{aligned}$$

Since for all $\varepsilon \in [0, 1]$

$$|I_2| \le \frac{f(0)}{\gamma}\int_0^{\gamma}(1-e^{-u^2})\,du \le f(0)(1-e^{-\gamma^2}),$$

the quantity $I_1 + I_2$ tends to zero as $\gamma \to 0$, which proves the lemma.

We now dwell on the case of any $n \ge 1$. Here, having obtained (2), we introduce by analogy with (4) the function

$$\omega_\varepsilon(t) = \frac{1}{(\sqrt{2\pi})^n}\int_{-\infty}^{t_1}\cdots\int_{-\infty}^{t_n} \hat{f}_\varepsilon(v)\,dv_1\cdots dv_n$$

for any $t = \{t_1, \dots, t_n\} \in E_n$. This function $\omega_\varepsilon(t)$ is a monotone function of $t$ in the sense that $\omega_\varepsilon(t) \ge \omega_\varepsilon(t')$ if $t_1 \ge t_1', \dots, t_n \ge t_n'$. Instead of (5) we now have the representation

$$f(u)e^{-\varepsilon|u|^2} = \int_{E_n} e^{i(u,t)}\,d\omega_\varepsilon(t).$$

Helly's selection theorem is used next; here it says that there exist a sequence $\varepsilon_i \to 0$ and a monotone function $\omega(t)$ such that $\lim_{i\to\infty}\omega_{\varepsilon_i}(t) = \omega(t)$. The lemma also carries over to the case of any $n$. In this case $(*)$ is replaced by the relation

$$\int_{E_n}\left\{1-\prod_{k=1}^{n}\frac{\sin\gamma t_k}{\gamma t_k}\right\}d\omega_\varepsilon(t) = f(0) - \frac{1}{(2\gamma)^n}\int_{Q_n(2\gamma)} f(u)e^{-\varepsilon|u|^2}\,du.$$

The final formulation of the theorem is as follows.

BOCHNER-KHINTCHINE THEOREM. *If $f$ is a continuous positive definite function on $E_n$, then there exists a bounded monotone function $\omega(t)$ of $t$ such that everywhere in $E_n$*

$$f(x) = \int_{E_n} e^{i(x,t)}\, d\omega(t). \tag{6}$$

*Conversely, to every such function $\omega(t)$ there corresponds a continuous positive definite function given by* (6).

The second part of the theorem is trivial.

**3.** The next result is an essential supplement of the Bochner-Khintchine theorem.

THEOREM. *For a given continuous positive definite function $f(x)$ the monotone function $\omega$ in the representation* (6) *is uniquely determined at its points of continuity, to within an additive constant.*

We confine ourselves to proving this for $n = 1$. Take the continuous function $\Phi(t; a, b, \varepsilon)$ defined by

$$\Phi(t; a, b, \varepsilon) = \begin{cases} 1 & (a \le t \le b), \\ 0 & (t > b + \varepsilon \text{ or } t < a - \varepsilon), \end{cases}$$

and linear in the intervals $(a - \varepsilon, a)$ and $(b, b + \varepsilon)$, where $\varepsilon > 0$. The Fourier transform $\hat{\Phi} = \Psi$ of $\Phi$ belongs to $L^1$, as is easily seen by a simple computation. The formula

$$f(x) = \int_{-\infty}^{\infty} e^{ixt}\, d\omega(t)$$

implies that

$$\begin{aligned} &\frac{1}{\sqrt{2\pi}} \int_{-\infty}^{\infty} f(x)\Psi(x; a, b, \varepsilon)\, dx \\ &\quad = \int_{-\infty}^{\infty} d\omega(t) \cdot \frac{1}{\sqrt{2\pi}} \int_{-\infty}^{\infty} e^{ixt}\Psi(x; a, b, \varepsilon)\, dx = \int_{-\infty}^{\infty} \Phi(t; a, b, \varepsilon)\, d\omega(t) \\ &\quad = \int_{a-\varepsilon}^{a} \frac{t - a + \varepsilon}{\varepsilon}\, d\omega(t) + \int_a^b d\omega(t) + \int_b^{b+\varepsilon} \frac{t - b - \varepsilon}{-\varepsilon}\, d\omega(t). \end{aligned}$$

Assume that $a$ and $b$ are points of continuity of $\omega(t)$. Then the middle term of the right-hand side is equal to $\omega(b) - \omega(a)$, and the end terms tends to 0 as $\varepsilon \to 0$. Consequently,

$$\omega(b) - \omega(a) = \lim_{\varepsilon \to 0} \frac{1}{\sqrt{2\pi}} \cdot \int_{-\infty}^{\infty} f(x)\Psi(x; a, b, \varepsilon)\, dx. \tag{7}$$

Since $\omega(b) - \omega(a)$ is uniquely determined by $f$ for any points of continuity of $\omega(t)$, the theorem is proved.

**4.** Problems. 1. Prove that along with (7) the inversion formula can be represented in the form

$$\omega(b) - \omega(a) = \lim_{T \to \infty} \frac{1}{2\pi i} \int_{-T}^{T} f(x) \frac{e^{-iax} - e^{-ibx}}{x} \, dx \tag{$7_1$}$$

(this inversion formula is due to Lévy and Bernstein).

2. Generalize (7) and ($7_1$) to the case of any $n > 1$.

CHAPTER 9

# The Hankel Transform

**1.** This chapter is devoted to the construction of an integral operator whose kernel is a Bessel function. This operator is called the *Hankel transform.* We write the Bessel function $J_\nu(x)$ $(\nu > -1)$ as a series:

$$J_\nu(x) = \sum_{k=0}^{\infty}(-1)^k \left(\frac{x}{2}\right)^{\nu+2k} /\Gamma(\nu + k + 1) \cdot k!. \tag{1}$$

The expansion of the so-called *modified Bessel function*

$$I_\nu(x) = \sum_{k=0}^{\infty} \frac{(x/2)^{\nu+2k}}{\Gamma(\nu + k + 1) \cdot k!}$$

is a majorant of the previous series on the half-line $x > 0$. We need the following inequality:

$$I_\nu(x) \le \frac{1}{\Gamma(\nu + 1)} \cdot \left(\frac{x}{2}\right)^{\nu} + \left(\frac{x}{2}\right)^{\nu+2} \cosh x \qquad (x \ge 0).$$

To prove it we represent $I_\nu(x)$ in the form

$$I_\nu(x) = \frac{1}{\Gamma(\nu + 1)} \cdot \left(\frac{x}{2}\right)^{\nu} + \left(\frac{x}{2}\right)^{\nu+2} \sum_{k=0}^{\infty} \left(\frac{x}{2}\right)^{2k} /\Gamma(\nu + k + 2) \cdot \Gamma(k + 2).$$

It now remains to verify that for $k \ge 0$

$$2^{2k} \cdot \Gamma(\nu + k + 2) \cdot \Gamma(k + 2) \ge (2k)!.$$

This inequality is obvious for $k = 0$, and the rest can be checked by induction. The following integral is needed for our constructions:

$$\int_0^\infty e^{-r^2/2} r^{\nu+1} J_\nu(tr)\, dr.$$

To find it we represent $J_\nu(tr)$ in the form (1). Thanks to an estimate of $I_\nu(tr)$, we then find that termwise integration can be used to compute the integral. Omitting the simple transformations, we write the final result:

$$\int_0^\infty e^{-r^2/2} r^{\nu+1} J_\nu(tr)\, dr = e^{-t^2/2} t^\nu,$$

which is conveniently represented in the form

$$\int_0^\infty e^{-r^2/2} r^{\nu+1/2} \sqrt{tr} J_\nu(tr)\, dr = e^{-t^2/2} t^{\nu+1/2}. \tag{2}$$

This formula expresses the fact that $e^{-t^2/2}t^{\nu+1/2}$ is an eigenfunction of the integral operator with kernel $\sqrt{tr}\cdot J_\nu(tr)$. The reader should consider the case of (2) when $\nu = -1/2$. Here it should be recalled that

$$J_{-1/2}(x) = \sqrt{2/\pi x}\cos x.$$

**2.** Consider formula (2), and in it let

$$r = \sqrt{\alpha}\, y, \quad t = \frac{1}{\sqrt{\alpha}} x, \qquad \alpha > 0.$$

We get the relation

$$\int_0^\infty e^{-\alpha y^2/2} \alpha^{(\nu+1)/2} y^{\nu+1} J_\nu(xy)\, dy = e^{-x^2/2\alpha} \cdot \alpha^{-\nu/2} \cdot x^\nu.$$

Setting $\beta = 1/\alpha$, we find that

$$\int_0^\infty e^{-\alpha y^2/2} \alpha^{(\nu+1)/2} y^{\nu+1/2} \sqrt{xy} J_\nu(xy)\, dy = e^{-\beta x^2/2} \cdot \beta^{(\nu+1)/2} \cdot x^{\nu+1/2}. \tag{3}$$

Observe now that $\alpha\partial/\partial\alpha = -\beta\partial/\partial\beta$, and apply the operator $(\alpha\partial/\partial\alpha)^k$ to the left-hand side of (3) and $(-\beta\partial/\partial\beta)^k$ to the right-hand side, setting $\alpha = \beta = 1$ in the result.

We then obtain the identity

$$\int_0^\infty e^{-y^2/2} y^{\nu+1/2} T_k(y^2) \sqrt{xy} J_\nu(xy)\, dy = (-1)^k e^{-x^2/2} x^{\nu+1/2} T_k(x^2), \tag{4}$$

where $T_k(z)$ is a $k$th-degree polynomial in $z$. In particular,

$$T_0(z) = 1, \quad T_1(z) = z - (\nu + 1), \quad T_2(z) = z^2 - 2(\nu+2)z + (1+\nu)^2.$$

*Thus, the expressions*

$$e^{-x^2/2} \cdot x^{\nu+1/2} \cdot T_{2k}(x^2) \qquad (k = 0, 1, 2, \dots) \tag{5}$$

*are eigenfunctions of the operator* $\mathscr{H}$ *defined by*

$$(\mathscr{H} f)(x) = \int_0^\infty f(y) \sqrt{xy} J_\nu(xy)\, dy$$

*that belong to the eigenvalue* 1, *and, similarly, the expressions*

$$e^{-x^2/2} \cdot x^{\nu+1/2} \cdot T_{2k+1}(x^2) \qquad (k = 0, 1, 2, \dots) \tag{6}$$

*belong to the eigenvalue* $-1$. *Since* $\mathscr{H}$ *is symmetric, each of the functions* (5) *is orthogonal to each of the functions* (6).

It should be mentioned that all the integrals in (4) are absolutely convergent, since the functions (5) and (6) belong both to $L^1(0,\infty)$ and to $L^2(0,\infty)$. The functions (5) and the functions (6) can be separately orthogonalized and normalized (it is more convenient for us to do this by 1/2). Combining them, we get the sequence of functions

$$e^{-x^2/2}x^{\nu+1/2}\tilde{L}_k^{(\nu)}(x^2) \qquad (k=0,1,2,\dots), \tag{7}$$

where $\tilde{L}_k^\nu$ is a polynomial of precise degree $k$. The orthonormality is expressed by the equality

$$\int_0^\infty e^{-x^2}x^{2\nu+1}\tilde{L}_k^{(\nu)}(x^2)\tilde{L}_m^{(\nu)}(x^2)\,dx = \frac{1}{2}\delta_{km}. \tag{8}$$

Note that each of the functions (7) is an eigenfunction of our operator belonging to the eigenvalue $(-1)^k$. If we make the substitution $x^2 = t$ in (7), we get instead of (8)

$$\int_0^\infty e^{-t}t^\nu \tilde{L}_k^{(\nu)}(t)\tilde{L}_m^{(\nu)}(t)\,dt = \delta_{km},$$

where the $\tilde{L}_k^{(\nu)}(t)$ are the normalized *Sonin-Laguerre polynomials*. They play an important role in analysis.

One often considers nonnormalized polynomials determined by

$$e^{-x}x^\nu L_n^{(\nu)}(x) = \frac{1}{n!}\frac{d^n}{dx^n}(e^{-x}x^{n+\nu}).$$

**3.** Let us show that the system of Sonin-Laguerre polynomials is *complete*. This means that *if a measurable function $f$ satisfies the conditions*

1) $\displaystyle\int_0^\infty e^{-z}\cdot x^\nu\cdot|f(x)|^2\,dx<\infty,$

2) $\displaystyle\int_0^\infty e^{-x}\cdot x^\nu\cdot f(x)\cdot L_n^{(\nu)}(x)\,dx = 0 \qquad (n=0,1,2,\dots),$

*then $f(x)=0$ a.e.* The conditions 2) are equivalent to

2′) $\displaystyle\int_0^\infty e^{-x}x^\nu f(x)x^n\,dx = 0 \qquad (n=0,1,2,\dots).$

Consider the function

$$g(\zeta) = \frac{1}{\sqrt{2\pi}}\int_0^\infty e^{-x}\cdot x^\nu\cdot f(x)\cdot e^{-ix\zeta}\,dx,$$

where $\zeta = \xi + i\eta$ ($|\eta| < 1/4$). In this strip $|e^{-ix\zeta}| < e^{x/4}$, and, since $e^{-x/2}\cdot x^{\nu/2}\cdot f(x)\in L^2(0,\infty)$ and $e^{-x/4}x^{\nu/2}x^n \in L^2(0,\infty)$ for any $n\ge 0$, the integrand and its products by $x^n$ ($n=0,1,\dots$) belong to $L^1(0,\infty)$ for any point $\zeta$ in the strip. This impies that $g(\zeta)$ is analytic in this strip, and its derivatives are

$$g^{(n)}(\zeta) = \frac{1}{\sqrt{2\pi}} \int_0^\infty e^{-x} x^\nu (-ix)^n f(x) e^{-ix\zeta}\, dx.$$

By condition 2′) $g^{(n)}(0) = 0$ for $n = 0, 1, \dots$. But in this case the analytic function $g(\zeta)$ is identically zero in the whole strip. In particular, it is equal to zero in the real axis, i.e., for $\eta = 0$, Thus,

$$0 = g(\xi) = \frac{1}{\sqrt{2\pi}} \int_0^\infty e^{-x} x^\nu f(x) e^{-ix\xi}\, dx.$$

The right-hand side is $\hat{F}(\xi)$, where

$$F(x) = \begin{cases} e^{-x} x^\nu f(x) & (x > 0), \\ 0 & (x < 0). \end{cases}$$

It follows from the uniqueness theorem that $F(x) = 0$ a.e.

**4.** In this section we consider for a fixed $\nu > -1$ the *Hankel operator* $\mathscr{H}_\nu = \mathscr{H}$ defined by

$$(\mathscr{H} f)(x) = \int_0^\infty f(y) \sqrt{xy}\, J_\nu(xy)\, dy,$$

which makes sense directly if $f \in L^1(0, \infty)$ and $\int_0^1 |f(y)| y^{\nu+1/2}\, dy < \infty$.(22) However, we assume that $f \in L^2(0, \infty)$. Under this assumption $f$ has an orthogonal series expansion in the functions (7):

$$f(x) \overset{L^2}{=} \sum_{k=0}^\infty a_k e^{-x^2/2} x^{\nu+1/2} v \tilde{L}_k^{(\nu)}(x^2), \tag{9}$$

where

$$a_k = 2 \int_0^\infty e^{-y^2/2} y^{\nu+1/2} \tilde{L}_k^{(\nu)}(y^2) f(y)\, dy.$$

The expansion (9) should be understood in the $L^2(0, \infty)$-metric. This means that if

$$f_n(x) = \sum_{k=0}^n a_k e^{-x^2/2} x^{\nu+1/2} \tilde{L}_k^{(\nu)}(x^2),$$

then

$$\lim_{n\to\infty} \|f - f_n\|_2 = 0.$$

But, by the considerations in §2,

$$(\mathscr{H} f_n)(x) = \sum_{k=0}^n (-1)^k a_k e^{-x^2/2} x^{\nu+1/2} \tilde{L}_k^{(\nu)}(x^2).$$

(22)Note that the second condition is satisfied if $f \in L^2(0, 1)$.

Therefore,

$$\|\mathscr{H} f_n - \mathscr{H} f_m\|_2 = \|f_n - f_m\|_2 \quad \text{and} \quad \|\mathscr{H} f_n\|_2 = \|f_n\|_2.$$

From this it is easy to see that the Hankel operator is well-defined on any element $f \in L^2(0, \infty)$ by the formula $\mathscr{H} f = \operatorname{l.i.m.}_{n\to\infty} \mathscr{H} f_n$, i.e.,

$$(\mathscr{H} f)(x) \overset{L^2}{=} \sum_{k=1}^{\infty} (-1)^k a_k e^{-x^2/2} x^{\nu+1/2} \tilde{L}_k^{(\nu)}(x^2)$$

if $f$ is given by (9). This operator is unitary and selfadjoint, and can also be defined by

$$(\mathscr{H} f)(x) = \operatorname*{l.i.m.}_{A\to\infty} \int_0^A f(y)\sqrt{xy}\, J_\nu(xy)\, dy.$$

The proof of these simple facts is left to the reader.

5. PROBLEMS. 1. Prove that the eigenfunctions of the Hankel operator $v_k = e^{-x^2/2} x^{\nu+1/2} L_k^{(\nu)}(x^2)$ are eigenfunctions of the differential operator

$$-\frac{d^2}{dx^2} + \left(x^2 + \frac{\nu^2 - 1/4}{x^2}\right) \qquad (0 < x < \infty; v > -1), \qquad (*)$$

namely,

$$-\frac{d^2 v_k}{dx^2} + \left(x^2 + \frac{\nu^2 - 1/4}{x^2}\right) v_k = \lambda_k v_k \qquad (0 < x < \infty),$$

where $\lambda_k = 4k + 2(\nu + 1)$; the spectrum of the operator $(*)$ is exhausted by the collection $\lambda_k$ $(k = 0, 1, \dots)$.

2. Prove the recursion formula

$$kL_k^{(\nu)}(x) = (-x + 2k + \nu - 1)L_{k-1}^{(\nu)} - (k + \nu - 1)L_{k-2}^{(\nu)}(x) \qquad (k = 2, 3, \dots).$$

CHAPTER 10

# Orthogonal Polynomials and the Moment Problem

**1.** The special orthogonal systems of polynomials encountered above, namely, the Tchebycheff-Hermite polynomials, the Sonin-Laguerre polynomials, and the Gegenbauer polynomials (the latter are a special case of the Jacobi polynomials) are called *classical systems*. In this chapter we consider a general approach to orthogonal polynomials and certain important results to which this approach leads.

We first take a monotonically increasing (more precisely, nondecreasing) bounded function $\sigma(u)$ $(-\infty < u < \infty)$ such that all the Stieltjes integrals

$$s_k = \int_{-\infty}^{\infty} u^k \, d\sigma(u) \qquad (k = 0, 1, \dots)$$

are finite; these integrals are commonly called the *moments* of the distribution $\sigma(u)$. To exclude some degenerate cases not of interest to us we assume that $\sigma(u)$ has an infinite set of points of increase. Under this condition the strict inequality

$$\int_{-\infty}^{\infty} R(u) \, d\sigma(u) > 0 \tag{1}$$

holds for any polynomial $R(u)$ that is nonnegative on the whole line and not identically equal to zero. We can now pose the problem of constructing an infinite sequence of polynomials $P_n(u)$ ($n = 0, 1, \dots$ is the degree of the polynomial) such that

$$\int_{-\infty}^{\infty} P_n(u) P_k(u) \, d\sigma(u) = \delta_{nk}; \tag{2}$$

such polynomials are said to be *orthonormal with respect to* $\sigma$.

Known procedures in linear algebra make it possible to construct these polynomials successively. It also follows from these procedures that the polynomials are uniquely determined to within a sign. It is convenient for us to write the general form of the orthonormal polynomials as determinants. After these formulas are written, their proofs will not present any difficulties. Here is the general formula:

$$P_n(u) = C_n \begin{vmatrix} s_0 & s_1 & \dots & s_n \\ s_1 & s_2 & \dots & s_{n+1} \\ \dots & \dots & \dots & \dots \\ s_{n-1} & s_n & \dots & s_{2n-1} \\ 1 & u & \dots & u^n \end{vmatrix} \qquad (n = 1, 2, \dots), \tag{3}$$

where $C_n$ is a constant depending on $n$. It suffices to prove that $P_n(u)$ is orthogonal to $u^k$ for $k = 0, \dots, n-1$. To do this we multiply $P_n(u)$ by $u^k$:

$$P_n(u)u^k = C_n \begin{vmatrix} s_0 & s_1 & \dots & s_n \\ s_1 & s_2 & \dots & s_{n+1} \\ \dots & \dots & \dots & \dots \\ s_{n-1} & s_n & \dots & s_{2n-1} \\ u^k & u^{k+1} & \dots & u^{k+n} \end{vmatrix}.$$

We must now integrate with respect to the measure $d\sigma$. For this it is necessary to carry out the integration in the last row of the determinant, after which we get in it the numbers $s_k, \dots, s_{n+k}$, which coincide with the elements of the $(k+1)$st row. Therefore, the integral is equal to 0. It is immediately clear that the function $\sigma(u)$ itself is not needed in this construction; only its moments are needed. We observe also that not even the integration was necessary for the construction. We could have proceeded formally, introducing a functional linear in the algebraic sense (i.e., additive and homogeneous), traditionally denoted by $\mathfrak{S}$, that is determined on the collection of all polynomials in $u$ by the relation

$$\mathfrak{S}\{u^k\} = s_k \qquad (k = 0, 1, \dots).$$

In this case (2) is written in the form

$$\mathfrak{S}\{P_n(u)P_k(u)\} = \delta_{nk}. \tag{4}$$

However, it is still necessary to explain what inequality (1), valid for any polynomial $R(u)$ such that $R(u) \geq 0$ $(-\infty < u < \infty)$ and $R(u) \not\equiv 0$, now reduces to. The requirement (1) followed from the assumption that $\sigma(u)$ has an infinite set of points of increase. But $\sigma(u)$ now does not take part in our considerations, and we must determine what conditions must be imposed on the sequence $\{s_k\}_0^\infty$ of moments in order for every polynomial $R(u) \geq 0$ $(-\infty < u < \infty)$ and not identically 0 to satisfy the inequality $\mathfrak{S}\{R(u)\} > 0$. We prove that this requirement holds if and only if the quadratic form([23])

$$\sum_{i,k=0}^{n} s_{i+k}x_i x_k \tag{5}$$

([23]) Such forms are called *Hankel forms*.

is *positive-definite* for any $n$, i.e., it has a positive value for any nontrivial collection of real numbers $x_k$. If $R$ is taken to be

$$R(u) = \left[\sum_{k=0}^{n} x_k u^k\right]^2 \equiv [A(u)]^2,$$

where all the $x_k$ are real, then we get that

$$\mathfrak{S}\{R(u)\} = \sum_{i,k=0}^{n} x_i x_k \mathfrak{S}\{u^{i+k}\} = \sum_{i,k=0}^{n} s_{i+k} x_i x_k.$$

Because of this circumstance, our assertion follows from the next proposition.

LEMMA. *A polynomial $R(u)$ is nonnegative on the whole axis if and only if it can be represented in the form $R(u) = [A(u)]^2 + [B(u)]^2$, where $A(u)$ and $B(u)$ are two real polynomials.*

We need only prove the necessity of this representation. Assume that $R(u)$ satisfies the condition of the lemma. The real roots of this polynomial must therefore be of even multiplicity, and the complex roots must be pairwise conjugate. Consequently,

$$R(u) = c\prod_{k=1}^{m}(u-x_k)^2\prod_{l=1}^{n}[(u-\alpha_l)^2+\beta_l^2] \qquad (c>0),$$

where the factors in each product need not be distinct, so that the existence of real roots of even multiplicity $> 2$ and of complex roots of multiplicity $> 1$ is not excluded. It follows from the representation that

$$R(u) = c\left[\prod_{k=1}^{m}(u-x_k)\right]^2 \cdot \left|\prod_{l=1}^{n}(u-\alpha_l-i\beta_l)\right|^2.$$

Setting

$$\sqrt{c}\prod_{k=1}^{m}(u-x_k)\prod_{l=1}^{n}(u-\alpha_l-i\beta_l) = A(u)+iB(u),$$

where the $A(u)$ and $B(u)$ are real polynomials, we get the desired representation:

$$R(u) = [A(u)]^2 + [B(u)]^2.$$

**2.** Accordingly, assume that the sequence $\{s_k\}_0^\infty$ is positive-definite (this is understood to mean that the quadratic form (5) is positive-definite for any $n$). We construct polynomials $P_k(u)$ $(k = 0, 1, 2, \dots)$ satisfying the conditions (4). These polynomials are said to be orthonormal with respect to the (positive-definite) sequence $\{s_k\}_0^\infty$. They have the form (3), where the constants $C_k$ are not yet determined. We express these constants in terms of the determinants

$$D_k = \begin{vmatrix} s_0 & s_1 & \dots & s_k \\ s_1 & s_2 & \dots & s_{k+1} \\ \dots & \dots & \dots & \dots \\ \dots & \dots & \dots & \dots \\ s_k & s_{k+1} & \dots & s_{2k} \end{vmatrix}, \qquad D_0 = s_0.$$

First of all, note that $P_n(u) = C_n D_{n-1} u^n + R_{n-1}(u)$, where $R_{n-1}(u)$ is a polynomial of degree $n-1$. Therefore, by the orthogonality condition,

$$1 = \mathfrak{S}\{P_n^2(u)\} = \mathfrak{S}\{P_n(u) C_n D_{n-1} u^n\} = C_n^2 D_{n-1} D_n,$$

so that $C_n = 1/\sqrt{D_{n-1}D_n}$, and hence

$$P_n(u) = \frac{1}{\sqrt{D_{n-1}D_n}} \begin{vmatrix} s_0 & s_1 & \dots & s_n \\ \dots & \dots & \dots & \dots \\ \dots & \dots & \dots & \dots \\ s_{n-1} & s_n & \dots & s_{2n-1} \\ 1 & u & \dots & u^n \end{vmatrix} \qquad (n = 1, 2, \dots);$$
$$P_0(u) = \frac{1}{\sqrt{s_0}} = \frac{1}{\sqrt{D_{-1}D_0}}, \tag{3'}$$

if we set $D_{-1} = 1$, after which the first formula in (3′) is true also for $n = 0$.

**3.** We prove some general properties of orthogonal polynomials.

PROPOSITION I. *An orthogonal polynomial $P_n(u)$ changes sign at $n$ points of the real axis. In particular, all its roots are real and simple.*

PROOF. If we assume that $P_n(u)$ changes sign only at $k < n$ points $x_1 < \dots < x_k$, then the product $P_n(u)(u - x_1)\cdots(u - x_k)$ does not change sign on the axis, and, since the leading coefficient of $P_n(u)$ is positive, the whole product is nonnegative on the real axis, and it is not identically zero. Hence, $\mathfrak{S}\{P_n(u)(u-x_1)\cdots(u-x_k)\} > 0$, but this contradicts the fact that $P_n(u)$ is an orthogonal polynomial of degree $n$ with respect to $\{s_k\}_0^\infty$.

PROPOSITION II. *Orthogonal polynomials are connected by a trinomial recursion formula* $(n = 1, 2, \dots)$

$$uP_n(u) = A_n P_{n+1}(u) + B_n P_n(u) + C_n P_{n-1}(u),$$

*where $A_n, B_n$, and $C_n$ are constants.*

PROOF. Since $P_k(u)$ is a polynomial of degree precisely $k$, the three numbers $A_n, B_n$, and $C_n$ can be determined (and uniquely so, of course) in such a way that the polynomial

$$uP_n(u) - A_n P_{n+1}(u) - B_n P_n(u) - C_n P_{n-1}(u) = R(u)$$

has degree at most $n-2$. If we now multiply both sides of this equality by $R(u)$ and apply the functional $\mathfrak{S}$ to both sides, then we get zero on the left-hand side because of the orthogonality property. Hence $\mathfrak{S}\{R^2(u)\} = 0$, and this implies that $R \equiv 0$, since $R$ is a real polynomial.

PROPOSITION III. *There exists a system of positive numbers $\mu_k^{(n)}$ ($k = 1, \ldots, n$) such that*

$$\mathfrak{S}\{T_{2n-1}\} = \sum_{k=1}^{n} \mu_k^{(n)} T_{2n-1}(\lambda_k^{(n)}) \tag{6}$$

*for any polynomial $T_{2n-1}(u)$ of degree at most $2n-1$, where the $\lambda_k^{(n)}$ are the roots of $P_n(u)$.*

REMARK. If the functional $\mathfrak{S}$ is a Stieltjes integral with respect to the measure $d\sigma(u)$ or is an ordinary Riemann integral, then the formula of Proposition III is usually called a *Gaussian quadrature formula.*

PROOF. We write the elementary identity

$$T_{2n-1}(u) = P_n(u) Q_{n-1}(u) + R_{n-1}(u), \tag{7}$$

where $Q_{n-1}$ is the quotient when $T_{2n-1}$ is divided by $P_n$, and $R_{n-1}$ is the remainder. Both these polynomials have degree $\leq n-1$. Applying $\mathfrak{S}$ to both sides, we get that $\mathfrak{S}\{T_{2n-1}(u)\} = \mathfrak{S}\{R_{n-1}(u)\}$. On the other hand, we find from the Lagrange interpolation formula with nodes $\lambda_k^{(n)}$ ($k = 1, \ldots, n$) that

$$R_{n-1}(u) = \sum_{k=1}^{n} \frac{P_n(u) R_{n-1}(\lambda_k^{(n)})}{(u - \lambda_k^{(n)}) \cdot P_n'(\lambda_k^{(n)})} = \sum_{k=1}^{n} \frac{P_n(u) T_{2n-1}(\lambda_k^{(n)})}{(u - \lambda_k^{(n)}) \cdot P_n'(\lambda_k^{(n)})},$$

since $T_{2n-1}(\lambda_k^{(n)}) = R_{n-1}(\lambda_k^{(n)})$ by (7). Setting

$$\mu_k^{(n)} = \mathfrak{S}\left\{ \frac{P_n(u)}{(u - \lambda_k^{(n)}) \cdot P_n'(\lambda_k^{(n)})} \right\} = \mathfrak{S}\left\{ \frac{P_n(u) - P_n(\lambda_k^{(n)})}{(u - \lambda_k^{(n)}) \cdot P_n'(\lambda_k^{(n)})} \right\},$$

we get (6). It remains to prove that $\mu_k^{(n)} > 0$. For this we take $T_{2n-1}(u)$ to be the polynomial

$$\left[\frac{P_n(u) - P_n(\lambda_k^{(n)})}{(u - \lambda_k^{(n)}) \cdot P_n'(\lambda_k^{(n)})}\right]^2,$$

which is equal to 1 for $u = \lambda_k^{(n)}$ and to 0 for $u = \lambda_i^{(n)}$ $(i \neq k)$, and which is positive at all other points of the $u$-axis. Applying the quadrature formula (6) to it, we get on the right-hand side the number $\mu_k^{(n)}$, and this implies that $\mu_k^{(n)} > 0$.

**4.** We now formulate the power moment problem, which has important significance in contemporary science and a very interesting and instructive history, going back to Tchebycheff.

THE MOMENT PROBLEM. *Let $\{s_k\}_0^\infty$ be a given positive-definite sequence. Find a nondecreasing function $\sigma(u)$ $(-\infty < u < \infty)$ such that*

$$s_k = \int_{-\infty}^{\infty} u^k \, d\sigma(u) \qquad (k = 0, 1, 2, \ldots).$$

We prove here that at least one of the functions required by the problem exists, but from the problems at the end of the chapter it will be clear that uniqueness does not always hold.([24]) A complete investigation of the moment problem goes outside the framework of this book.

REMARK. The moment problem can be regarded as the problem of extending a certain positive functional. Indeed, if the function $\sigma(u)$ is found, then the functional $\mathfrak{S}$ on the set of polynomials turns out to be extended to a certain collection of arbitrary continuous functions defined on the axis. Further, the property of positivity is preserved, at least in a weakened form, since the inequality

$$\int_{-\infty}^{\infty} f(u) \, d\sigma(u) \geq 0$$

follows in any case from $f(u) \geq 0$ $(-\infty < u < \infty)$.

The proof of the existence of a function $\sigma(u)$ is based on the quadrature formula (6). We first write this formula in another form. To do this we

([24]) Two solutions of the moment problem are not regarded as different if their difference is constant on the set of all points of continuity of this difference. If the moment problem has a unique solution in this sense, then it is called a *determinate* moment problem, and otherwise an *indeterminate* moment problem.

introduce the function $\sigma_n(u)$ defined by

$$\sigma_n(u) = \begin{cases} 0 & (-\infty < u \le \lambda_1^{(n)}), \\ \mu_1^{(n)} & (\lambda_1^{(n)} < u \le \lambda_2^{(n)}), \\ \mu_1^{(n)} + \mu_2^{(n)} & (\lambda_2^{(n)} < u \le \lambda_3^{(n)}), \\ \dots & \\ \mu_1^{(n)} + \mu_2^{(n)} + \cdots + \mu_n^{(n)} & (\lambda_n^{(n)} < u < \infty), \end{cases}$$

in other words, $\sigma_n(u)$ is a piecewise constant function with $n$ points of increase $\lambda_1^{(n)} < \lambda_2^{(n)} < \cdots < \lambda_n^{(n)}$ and with jumps $\sigma_n(\lambda_k^{(n)}+0) - \sigma_n(\lambda_k^{(n)}) = \mu_k^{(n)}$. For each $k \le 2n-1$ we can write the equality

$$s_k = \int_{-\infty}^{\infty} u^k \, d\sigma_n(u), \tag{8}$$

which is equivalent to the quadrature formula (6). In particular, (8) implies that $0 \le \sigma_n(u) \le s_0$, so that Helly's selection theorem is applicable to the sequence $\{\sigma_n(u)\}_{n=1}^{\infty}$.(25) According to this theorem, there exists a sequence $\{\sigma_{n_i}(u)\}_{i=1}^{\infty}$ converging at each point to some monotone function $\sigma(u)$, $0 \le \sigma(u) \le s_0$. We show that $\sigma(u)$ is a solution of the moment problem.

Let $k$ be a positive integer, and take an even number $2r > k$. For any $A, B > 0$ the equality

$$\begin{aligned} s_k - \int_{-A}^{B} u^k \, d\sigma_{n_i}(u) &= \int_{-\infty}^{-A} u^k \, d\sigma_{n_i}(u) + \int_{B}^{\infty} u^k \, d\sigma_{n_i}(u) \\ &= \int_{-\infty}^{-A} \frac{u^{2r}}{u^{2r-k}} \, d\sigma_{n_i}(u) + \int_{B}^{\infty} \frac{u^{2r}}{u^{2r-k}} \, d\sigma_{n_i}(u) \end{aligned}$$

implies that

$$s_k - \int_{-A}^{B} u^k \, d\sigma_{n_i}(u)| \le \frac{1}{C^{2r-k}} \left\{ \int_{-\infty}^{-A} u^{2r} \, d\sigma_{n_i}(u) + \int_{B}^{\infty} u^{2r} \, d\sigma_{n_i}(u) \right\},$$

where $C = \min\{A, B\}$. For $n_i > r$ we get from this and (8) that

$$\left| s_k - \int_{-A}^{B} u^k \, d\sigma_{n_i}(u) \right| \le \frac{s_{2r}}{C^{2r-k}}.$$

By Helly's limit theorem,(26)

$$\left| s_k - \int_{-A}^{B} u^k \, d\sigma(u) \right| \le \frac{s_{2r}}{C^{2r-k}}.$$

---

(25) See footnote (21) in Chapter 8.

(26) See footnote (21) in Chapter 8.

Letting $A, B \to \infty$, we get that for all $k$

$$s_k = \int_{-\infty}^{\infty} u^k \, d\sigma(u),$$

which is what we claimed.

**5. Problems.** (27) 1. Let $\sum_{n=1}^{\infty} 1/\sqrt[2n]{s_{2n}} = \infty$. Prove that the moment problem

$$s_k = \int_{-\infty}^{\infty} u^k \, d\sigma(u) \qquad (k = 0, 1, 2, \dots) \tag{$*$}$$

is determinate (Carleman).

2. Let $s_k = \int_{-\infty}^{\infty} u^k \, d\tau(u)$ $(k = 0, 1, 2, \dots)$, where $\tau(u)$ $(-\infty < u < \infty)$ is a nondecreasing function for which

$$\int_{-\infty}^{\infty} \frac{\ln \tau'(u)}{1+u^2} \, du > -\infty$$

(here $\tau'(u)$ is the derivative of the absolutely continuous part of $\tau(u)$). Prove that the sequence $\{s_k\}_0^{\infty}$ gives rise to an indeterminate moment problem $(*)$ (M. G. Kreĭn).

3. Prove that for

$$s_k = (2k+1)!\left(1 + \frac{(-1)^k}{\alpha^{2k+2}}\right) \qquad (k = 0, 1, 2, \dots),$$

where $\alpha > 1$, the moment problem $(*)$ is indeterminate.

*Hint.* Verify that the numbers $s_k$ considered here admit the representation

$$s_k = \int_{-\infty}^{\infty} u^k \varphi(u) \, du,$$

where

$$\varphi(u) = \begin{cases} \frac{1}{2} e^{-\sqrt{u}} & (u > 0), \\ \frac{1}{2} e^{-\alpha\sqrt{|u|}} & (u < 0). \end{cases}$$

(27) Concerning Problems 1 and 2 see N. I. Akhiezer, *The classical moment problem and some related questions in analysis*, Fizmatgiz, Moscow, 1961, supplements and problems for Chapter II; English transl., Oliver and Boyd, London, and Hafner, New York, 1965.

CHAPTER 11

# The Class $H_2^+$. The Paley-Wiener Theorem

**1.** In this chapter we consider certain classes of analytic functions of a single complex variable; the Fourier transform on $L^2$ plays a role in the study of these classes.

DEFINITION. The function $F(z)$ $(z = x + iy)$ belongs to the *Hardy class* $H_2^+$ if is is analytic in the half-plane $y > 0$ and satisfies the inequality

$$M^2 \stackrel{\text{def}}{=} \sup_{y>0} \int_{-\infty}^{\infty} |F(x+iy)|^2 \, dx < \infty.$$

LEMMA. *If $F(z) \in H_2^+$ $(z = x + iy)$, then for any $b > 0$ and any $y > b$*

$$F(z) = \frac{1}{2\pi i} \int_{-\infty}^{\infty} \frac{F(\xi + ib)\, d\xi}{\xi + ib - z}, \tag{1}$$

$$0 = \frac{1}{2\pi i} \int_{-\infty}^{\infty} \frac{F(\xi + ib)\, d\xi}{\xi + ib - \bar{z}}. \tag{1'}$$

PROOF. Take a rectangle with vertices $(-A, b), (A, b), (A, B), (-A, B)$, where $A > 0$ and $B > b$ are large enough that the point $z$ lies inside this rectangle. Then, by the Cauchy formula,

$$\begin{aligned} F(z) = {} & \frac{1}{2\pi i} \int_{-A}^{A} \frac{F(\xi + ib)}{\xi + ib - z}\, d\xi + \frac{1}{2\pi} \int_{b}^{B} \frac{F(A + i\eta)}{A + i\eta - z}\, d\eta \\ & + \frac{1}{2\pi i} \int_{A}^{-A} \frac{F(\xi + iB)}{\xi + iB - z}\, d\xi + \frac{1}{2\pi} \int_{B}^{b} \frac{F(-A + i\eta)}{-A + i\eta - z}\, d\eta, \end{aligned}$$

and if $z$ is replaced by $\bar{z}$ on the right-hand side, then the sum equals 0 instead of $F(z)$. We now write the identity

$$\int_{-A}^{A} \frac{F(\xi + ib)}{\xi + ib - z}\, d\xi = \int_{-\infty}^{\infty} \frac{F(\xi + ib)}{\xi + ib - z}\, d\xi - \int_{|\xi| > A} \frac{F(\xi + ib)}{\xi + ib - z}\, d\xi.$$

Further, by the Cauchy-Schwarz-Bunyakovskiĭ inequality,

$$\left|\int_{|\xi|>A}\right| \leq \sqrt{\int_{|\xi|>A} |F(\xi+ib)|^2\, d\xi}\sqrt{\int_{|\xi|>A} \frac{d\xi}{(\xi-x)^2+(y-b)^2}}$$

$$\leq M\sqrt{\int_{|t|>A-|x|} dt/t^2} = M\frac{\sqrt{2}}{\sqrt{A-|x|}} = O\left(\frac{1}{\sqrt{A}}\right) \qquad (A\to\infty).$$

Proceeding in exactly the same way with the integral

$$\int_{-A}^{A} \frac{F(\xi+iB)}{\xi+iB-z}\, d\xi,$$

we find that

$$F(z) = \frac{1}{2\pi i}\int_{-\infty}^{\infty} \frac{F(\xi+ib)}{\xi+ib-z}\, d\xi - \frac{1}{2\pi i}\int_{-\infty}^{\infty} \frac{F(\xi+iB)}{\xi+iB-z}\, d\xi + \mathfrak{A}, \qquad (\alpha)$$

where

$$\mathfrak{A} = O\left(\frac{1}{\sqrt{A}}\right) + \frac{1}{2\pi}\int_b^B \frac{F(A+i\eta)}{A+i\eta-z}\, d\eta - \frac{1}{2\pi}\int_b^B \frac{F(-A+i\eta)}{-A+i\eta-z}\, d\eta.$$

We multiply both sides by $dA$ and integrate with respect to $A$ from $C$ to $C+1$, where $C$ is sufficiently large. Since the left-hand side of $(\alpha)$ and the first two terms of the right-hand side do not depend on $A$, and since

$$\int_C^{C+1} O\left(\frac{1}{\sqrt{A}}\right) dA = O\left(\frac{1}{\sqrt{C}}\right),$$

we get the equality

$$F(z) = \frac{1}{2\pi i}\int_{-\infty}^{\infty} \frac{F(\xi+ib)}{\xi+ib-z}\, d\xi - \frac{1}{2\pi i}\int_{-\infty}^{\infty} \frac{F(\xi+iB)}{\xi+iB-z}\, d\xi + \mathfrak{B}, \qquad (\beta)$$

where

$$\begin{aligned}\mathfrak{B} &= O\left(\frac{1}{\sqrt{C}}\right) + \frac{1}{2\pi}\int_C^{C+1} dA\int_b^B \frac{F(A+i\eta)}{A+i\eta-z}\, d\eta \\ &\quad - \frac{1}{2\pi}\int_C^{C+1} dA\int_b^B \frac{F(-A+i\eta)}{-A+i\eta-z}\, d\eta \\ &= O\left(\frac{1}{\sqrt{C}}\right) + I_1 + I_2.\end{aligned}$$

We now use the estimates

$$|I_1| \leq \frac{1}{2\pi}\int_b^B d\eta\sqrt{\int_C^{C+1} |F(A+i\eta)|^2\, dA}\sqrt{\int_C^{C+1} \frac{dA}{(A-x)^2+(y-\eta)^2}}$$

$$\leq M\cdot\frac{B-b}{2\pi}\sqrt{\int_C^{C+1} \frac{dA}{(A-x)^2}} = O\left(\frac{1}{C}\right)$$

and $|I_2| = O(1/C)$. They imply that $\mathfrak{B} = O(1/\sqrt{C})$ as $C \to \infty$. Therefore, $(\beta)$ gives us

$$F(z) = \frac{1}{2\pi i}\int_{-\infty}^{\infty} \frac{F(\xi + ib)}{\xi + ib - z}\,d\xi - \frac{1}{2\pi i}\int_{-\infty}^{\infty} \frac{E(\xi + iB)}{\xi + iB - z}\,d\xi. \qquad (\gamma)$$

The second integral admits the estimate

$$\left|\frac{1}{2\pi i}\int_{-\infty}^{\infty} \frac{F(\xi + iB)}{\xi + iB - z}\,d\xi\right| \le \frac{M}{2\pi}\sqrt{\int_{-\infty}^{\infty} \frac{d\xi}{(\xi - x)^2 + (B - y)^2}}$$
$$= \frac{M}{2\sqrt{\pi}} \cdot \frac{1}{\sqrt{B - y}}.$$

Therefore, for any $z$ the second term on the right-hand side of $(\gamma)$ tends to 0 as $B \to \infty$, and hence (1) is proved. The same argument shows that (1′) is also true.

REMARK. If we denote by $H_2(\alpha, \beta)$ $(-\infty < \alpha < \beta < \infty)$ the class of functions $F(z)$ analytic in the strip $\alpha < y < \beta$ and satisfying

$$\sup_{\alpha < y < \beta} \int_{-\infty}^{\infty} |F(x + iy)|^2\,dx < \infty,$$

then in view of our considerations we can assert that for $\alpha < a < b < \beta$ and $y \in (a, b)$

$$F(z) = \frac{1}{2\pi i}\int_{-\infty}^{\infty} \frac{F(\xi + ia)}{\xi + ia - z}\,d\xi - \frac{1}{2\pi i}\int_{-\infty}^{\infty} \frac{F(\xi + ib)}{\xi + ib - z}\,d\xi. \qquad (2)$$

**2.** Passing to the connection between the class $H_2^+$ and the Fourier transform, we note that every function $F(z)$ of the form

$$F(z) = \frac{1}{\sqrt{2\pi}}\int_0^{\infty} e^{izt} f(t)\,dt,$$

where $f \in L^2(0, \infty)$, belongs to $H_2^+$. Indeed, this function is continuous and has a finite derivative at each point $z$ of the upper half-plane; and, on the other hand, it follows from the Parseval equality that for $y > 0$

$$\int_{-\infty}^{\infty} |F(x + iy)|^2\,dx = \int_0^{\infty} e^{-2yt}|f(t)|^2\,dt \le \int_0^{\infty} |f(t)|^2\,dt = M^2.$$

It turns out that the class $H_2^+$ is exhausted by these functions, i.e., we have

THEOREM 1. *For every function $F(z) \in H_2^+$ there exists a function $f \in L^2(0, \infty)$ such that for any $y > 0$*

$$F(z) = \frac{1}{\sqrt{2\pi}}\int_0^{\infty} e^{izt} f(t)\,dt \qquad (z = x + iy). \qquad (3)$$

PROOF. We take the representation (1) proved in the lemma and observe that for $y > b$

$$\frac{1}{\sqrt{2\pi}} \cdot \frac{1}{i(\xi + ib - z)} = \frac{1}{\sqrt{2\pi}} \int_0^\infty e^{-it(\xi + ib - z)}\, dt;$$

in other words, the function

$$\frac{1}{\sqrt{2\pi}} \cdot \frac{1}{i(\xi + ib - z)}$$

of $\xi$ is the Fourier transform of the function

$$\omega(t) = \begin{cases} e^{it(z-ib)} & (t > 0), \\ 0 & (t \leq 0). \end{cases}$$

Therefore, the representation (1) can be written in the form

$$F(z) = \frac{1}{\sqrt{2\pi}} \int_{-\infty}^{\infty} \hat{\omega}(\xi) F(\xi + ib)\, d\xi.$$

Denote by $f_b(t)$ the Fourier transform of the function $F(\xi + ib) \in L^2$. Then, by the Parseval equality,[28]

$$F(z) = \frac{1}{\sqrt{2\pi}} \int_{-\infty}^{\infty} \omega(t) f_b(t)\, dt = \frac{1}{\sqrt{2\pi}} \int_0^\infty e^{it(z-ib)} f_b(t)\, dt.$$

This is a certain representation of $F(z)$ for $y > b > 0$. Now take a positive $\beta < b$. Then we get the analogous representation

$$F(z) = \frac{1}{\sqrt{2\pi}} \int_0^\infty e^{it(z-i\beta)} f_\beta(t)\, dt,$$

true for $y > \beta$. We see that for $y > b$ there are two representations

$$F(x + iy) = \frac{1}{\sqrt{2\pi}} \int_0^\infty e^{itx} e^{t(b-y)} f_b(t)\, dt = \frac{1}{\sqrt{2\pi}} \int_0^\infty e^{itx} e^{t(\beta - y)} f_\beta(t)\, dt.$$

Both the functions

$$e^{t(b-y)} \cdot f_b(t), \quad e^{t(\beta - y)} \cdot f_\beta(t) \qquad (t > 0),$$

extended by zero for $t < 0$, belong to $L^1(-\infty, \infty)$ for any $y > b$, since each of them is a product of two functions in $L^2(-\infty, \infty)$. By the uniqueness theorem, $e^{t\beta} f_\beta(t) = e^{tb} f_b(t)$ almost everywhere on the semi-axis $(0, \infty)$. Consequently, the function $f(t) = e^{tb} f_b(t)$ $(0 < t < \infty)$ does not depend on $b$, and our representation has the form

$$F(z) = \frac{1}{\sqrt{2\pi}} \int_0^\infty e^{itz} f(t)\, dt$$

(28) See also Problem 1 in Chapter 5.

for any $y > 0$. Since the left-hand side belongs to $L^2(-\infty, \infty)$ with respect to $x$ uniformly for all $y > 0$, it follows from the Parseval equality that

$$\int_0^\infty e^{-2ty}|f(t)|^2\,dt \leq M^2$$

for any $y > 0$. This now gives us that

$$\int_0^\infty |f(t)|^2\,dt \leq M^2 < \infty,$$

i.e., the theorem is proved.

3. THEOREM 2. *For every function $F(z) \in H_2^+$ there exists a function $g(x) \in L^2(-\infty, \infty)$ such that*

$$\lim_{y\searrow 0}\int_{-\infty}^\infty |F(x+iy) - g(x)|^2\,dx = 0.$$

*Further, the representations*

$$F(z) = \frac{1}{2\pi i}\int_{-\infty}^\infty \frac{g(\xi)}{\xi - z}\,d\xi, \quad 0 = \frac{1}{2\pi i}\int_{-\infty}^\infty \frac{g(\xi)}{\xi - \bar{z}}\,d\xi \qquad (\operatorname{Im} z > 0)$$

*hold for any $y > 0$.*

PROOF. Since $F(z) \in H_2^+$, there exists a function $f(t) \in L^2(0, \infty)$ such that for $y > 0$

$$F(z) = \frac{1}{\sqrt{2\pi}}\int_0^\infty f(t)e^{itz}\,dt.$$

We introduce the function

$$g(x) = \frac{1}{\sqrt{2\pi}}\operatorname*{l.\,i.\,m.}_{A\to\infty}\int_0^A f(t)e^{itx}\,dt.$$

This is the inverse Fourier-Plancherel transform of the function equal to $f(t)$ for $t > 0$ and to 0 for $t < 0$. Then for $y > 0$

$$\int_{-\infty}^\infty |F(x+iy) - g(x)|^2\,dx = \int_0^\infty |f(t)|^2(1 - e^{-ty})^2\,dt,$$

by the Plancherel equality. According to the Lebesgue limit theorem, the right-hand side tends to 0 as $y \searrow 0$. The first assertion of the theorem is thus proved.

We now turn to the lemma. With its help we conclude that for fixed $y$ and $0 < b < y$

$$F(z) = \frac{1}{2\pi i}\int_{-\infty}^\infty \frac{F(\xi+ib) - g(\xi)}{\xi + ib - z}\,d\xi + \frac{1}{2\pi i}\int_{-\infty}^\infty \frac{g(\xi)\,d\xi}{\xi + ib - z};$$

$$0 = \frac{1}{2\pi i}\int_{-\infty}^\infty \frac{F(\xi+ib) - g(\xi)}{\xi + ib - \bar{z}}\,d\xi + \frac{1}{2\pi i}\int_{-\infty}^\infty \frac{g(\xi)\,d\xi}{\xi + ib - \bar{z}}.$$

Since, by the Cauchy-Schwarz-Bunyakovskiĭ inequality,

$$\begin{aligned}\left|\int_{-\infty}^{\infty} \frac{F(\xi+ib)-g(\xi)}{\xi+ib-x\mp iy}\,d\xi\right|^2 &\leq \int_{-\infty}^{\infty} |F(\xi+ib)-g(\xi)|^2\,d\xi \int_{-\infty}^{\infty} \frac{d\xi}{(\xi-x)^2+(b\mp y)^2} \\ &= \pi(y\mp b)^{-1}\int_{-\infty}^{\infty} |F(\xi+ib)-g(\xi)|^2\,d\xi,\end{aligned}$$

the first terms in the formulas $(*)$ tend to zero as $b\to 0$ in view of what has already been proved.

To finish the proof of Theorem 2 it remains to observe that

$$\begin{aligned}&\left|\int_{-\infty}^{\infty} g(\xi)\left(\frac{1}{\xi+ib-x\mp iy}-\frac{1}{\xi-x\mp iy}\right)d\xi\right| \\ &\quad\leq \|g\|_2\sqrt{\int_{-\infty}^{\infty}\frac{b^2\,d\xi}{[(x-\xi)^2+y^2]\cdot[(x-\xi)^2+(y\mp b)^2]}} \\ &\quad\leq \frac{b\sqrt{\pi}}{\sqrt{y}(y\mp b)}\cdot\|g\|_2\to 0 \qquad (b\to 0).\end{aligned}$$

COROLLARY. *Assume the conditions of Theorem 2, and let $g(x)\in L^2$ be the function in the statement of this theorem. Then the representation*

$$F(z)=\frac{1}{\pi}\int_{-\infty}^{\infty}\frac{yg(\xi)}{(x-\xi)^2+y^2}\,d\xi \qquad (y>0)$$

*is valid, and*

$$\lim_{y\searrow 0} F(x+iy)=g(x)$$

*almost everywhere.*

REMARK. We can determine $F$ on the real axis by

$$F(x)=F(x+0i)=\lim_{y\searrow 0}F(x+iy)=g(x) \quad \text{a.e.}$$

Then Theorem 2 and its corollary show that $F(z)$ *is representable for* $y>0$ *both as a Cauchy integral and as a Poisson integral.*

PROOF OF THE COROLLARY. It follows from the second assertion of Theorem 2 that for $y>0$

$$F(z)=\frac{1}{2\pi i}\int_{-\infty}^{\infty} g(\xi)\left\{\frac{1}{\xi-z}-\frac{1}{\xi-\bar z}\right\}d\xi=\frac{1}{\pi}\int_{-\infty}^{\infty} d(\xi)\frac{y\,d\xi}{(x-\xi)^2+y^2}.$$

This is the first assertion of the corollary. Further, since $g\in L^2(-\infty,\infty)$, it follows that

$$\frac{g(x)}{(1+|x|)^2}\in L^1(-\infty,\infty).$$

On the other hand, the kernel

$$K(x) = \frac{1}{\pi} \cdot \frac{1}{1+x^2}$$

satisfies the condition

$$(1+|x|)^2 K(x) \leq M < \infty.$$

Therefore, the Bochner representation theorem is applicable here and gives us the second assertion of the corollary.

THEOREM 3. *If for* $y > 0$

$$F(z) = \frac{1}{2\pi i} \int_{-\infty}^{\infty} \frac{g(\xi)\,d\xi}{\xi - z}, \qquad 0 = \frac{1}{2\pi i} \int_{-\infty}^{\infty} \frac{g(\xi)\,d\xi}{\xi - \bar{z}},$$

*where* $g(\xi) \in L^2(-\infty, \infty)$, *then* $F(z) \in H_2^+$, *and*

$$\lim_{y \searrow 0} F(x+iy) = g(x) \quad a.e.$$

PROOF. For $y > 0$

$$F(x+iy) = F(z) = \frac{1}{\pi} \int_{-\infty}^{\infty} \frac{g(\xi) y\,d\xi}{(x-\xi)^2 + y^2} = \frac{1}{\pi} \int_{-\infty}^{\infty} \frac{g(x+t) y}{t^2 + y^2}\,dt.$$

Taking the norm with respect to $x$ in $L^2(-\infty, \infty)$, we get that

$$\sqrt{\int_{-\infty}^{\infty} |F(x+iy)|^2 dx} \leq \frac{1}{\pi} \int_{-\infty}^{\infty} \frac{y\,dt}{t^2+y^2} \sqrt{\int_{-\infty}^{\infty} |g(x+t)|^2\,dt} = \|g\|_2,$$

which implies the first assertion of the theorem. The second follows from the corollary.

**4.** In connection with the constructions of the present chapter we consider the so-called *Laplace transform*, which is used frequently in mathematical physics and theoretical electrical engineering.

This transform is defined by

$$\Phi(s) = \int_0^{\infty} e^{-st} \varphi(t)\,dt; \tag{4}$$

$\Phi$ is the transform of $\varphi$, and $\varphi$ is called the *determining function* with respect to $\Phi$; $s = \sigma + i\tau$ is a complex variable. Above we considered functions in the upper half-plane, but now $s$ runs through a right half-plane, more precisely, the half-plane $\sigma > c$. Various assumptions are made about $\varphi$. We assume that $e^{-ct}\varphi(t) \in L^2(0, \infty)$. In this case $\Phi(s)$ is regular in the plane $\sigma > c$, and, as is easy to verify by making the change $z = is$ in the formulas of this chapter, the function $\Phi(s)$ satisfies the relation

$$\sup_{\sigma > c} \int_{-\infty}^{\infty} |\Phi(\sigma + i\tau)|^2\,d\tau < \infty.$$

We can say that $\Phi(s) \in H_2(\sigma > c)$. It is just necessary to keep in mind that the coefficient $1/\sqrt{2\pi}$ in the definition of the transform is omitted here. If we represent (4) in the form

$$\Phi(\sigma + i\tau) = \int_0^\infty e^{-i\tau t} e^{-\sigma t} \varphi(t)\, dt, \tag{4'}$$

so that on the right-hand side we have the Fourier transform of the function $e^{-\sigma t}\varphi(t)$, which is in $L^2$ for $\sigma = c$ and in $L^1$ for $\sigma > c$, then, by the inversion of this formula (4′),

$$e^{-\sigma t}\varphi(t) = \frac{1}{2\pi}\int_{-\infty}^\infty \Phi(\sigma + i\tau) e^{i\tau t}\, d\tau \qquad (t > 0)$$

and

$$0 = \frac{1}{2\pi}\int_{-\infty}^\infty \Phi(\sigma + i\tau) e^{i\tau t}\, d\tau \qquad (t < 0).$$

The two formulas can be rewritten in the form

$$\operatorname*{l.\,i.\,m.}_{T\to\infty} \frac{1}{2\pi i}\int_{\sigma - iT}^{\sigma + iT} \Phi(s) e^{st}\, ds = \begin{cases} \varphi(t) & (t > 0), \\ 0 & (t < 0). \end{cases} \tag{5}$$

The Parseval equality

$$\frac{1}{2\pi}\int_{-\infty}^\infty |\Phi(\sigma + i\tau)|^2\, d\tau = \int_0^\infty e^{-2ct} |\varphi(t)|^2\, dt \tag{6}$$

must still be adjoined to this.

**5.** To conclude this chapter we present an important theorem on entire functions of exponential type due to Paley and Wiener. Recall that an entire function $g(z)$ is called an *entire function of exponential type* $\leq A$ if for any $\alpha > 0$ there is a constant $B = B(\alpha)$ such that for any $z$

$$|g(z)| \leq B e^{(A+\alpha)|z|}.$$

THEOREM (PALEY-WEINER). *The class of entire functions $g(z)$ of exponential type $\leq A$ for which*

$$\int_{-\infty}^\infty |g(x)|^2\, dx < \infty$$

*coincides with the class of functions representable in the form*

$$g(z) = \frac{1}{\sqrt{2\pi}}\int_{-A}^A \varphi(t) e^{itz}\, dt,$$

*where $\varphi(t) \in L^2(-A, A)$.*

PROOF. If $g(z)$ admits such a representation, then for any $z$

$$|g(z)| \leq \frac{1}{\sqrt{2\pi}} e^{A|y|} \int_{-A}^\infty |\varphi(t)|\, dt,$$

and one part of the theorem is proved.

Suppose now that $g(z)$ is a given entire function of exponential type $\leq A$ such that $g(x) \in L^2(-\infty, \infty)$. According to the results of Chapter 5, §4,

$$g(x) = \frac{1}{\sqrt{2\pi}} \frac{d}{dx} \int_{-\infty}^{\infty} \frac{e^{itx} - 1}{it} \varphi(t)\, dt$$

almost everywhere, where $\varphi(t) \in L^2(-\infty, \infty)$. Therefore, for any $\varepsilon > 0$

$$\frac{1}{\varepsilon} \int_x^{x+\varepsilon} g(\xi)\, d\xi = \frac{1}{\sqrt{2\pi}} \int_{-\infty}^{\infty} \frac{e^{i\varepsilon t} - 1}{i\varepsilon t} \varphi(t) e^{itx}\, dt. \tag{7}$$

We introduce the entire function

$$G(z) = G(z; A', \varepsilon) = e^{iA'z} \frac{1}{\varepsilon} \int_z^{z+\varepsilon} g(\zeta)\, d\zeta,$$

which depends on the parameters $A' > A$ and $\varepsilon > 0$. On the real axis

$$|G(x)| \leq \frac{1}{\varepsilon} \int_x^{x+\varepsilon} |g(\xi)|\, d\xi = \frac{1}{\varepsilon} \int_0^{\varepsilon} |g(x+t)|\, dt, \tag{8}$$

which implies that

$$\begin{aligned} \int_{-\infty}^{\infty} |G(x)|^2\, dx &\leq \int_{-\infty}^{\infty} dx \frac{1}{\varepsilon^2} \left[ \int_0^{\varepsilon} |g(x+t)|\, dt \right]^2 \\ &\leq \int_{-\infty}^{\infty} dx \frac{1}{\varepsilon} \int_0^{\varepsilon} |g(x+t)|^2\, dt = \frac{1}{\varepsilon} \int_0^{\varepsilon} \|g\|_2^2 = \|g\|_2^2, \end{aligned}$$

i.e., $G(x) \in L^2(-\infty, \infty)$ and $\|G\|_2 \leq \|g\|_2$. Moreover, it follows from (8) that

$$|G(x)| \leq \frac{1}{\sqrt{\varepsilon}} \|g\|_2. \tag{9}$$

By the definition of the type of a function, for any $A'' > A$ there is a constant $B$ such that

$$|G(iy)| \leq e^{-A'y} \frac{1}{\varepsilon} \int_0^{\varepsilon} |g(iy + \xi)|\, d\xi \leq e^{-A'y} B e^{A''|iy+\varepsilon|}$$

for all $y$. Choose $A''$ such that $A < A'' < A'$ and let $\delta = A' - A''$. Then for $y > 0$ we have the estimate $|G(iy)| \leq Me^{-\delta y}$, where the constant $M$ depends on $\varepsilon$ and $A''$. Considering (9) and increasing $M$ if necessary, we can assume that the inequality $|G(z)e^{-i\delta z}| \leq M$ holds at the points of the real axis and at the points of the plane with $x = 0$ and $y > 0$. But the Phragmén-Lindelöf theorem,([29]) applied first to the first quadrant and next to the second quadrant, then gives us that this inequality is valid in

([29])See, for example, E. C. Titchmarsh, *The theory of functions*, 2nd ed., Clarendon Press, Oxford, 1939, Chapter 5.

the whole upper half-plane. It follows from Problem 2 in this chapter (see below) that $G(z) \in H_2^+$, and by Theorem 1

$$G(z) = \frac{1}{\sqrt{2\pi}} \int_0^\infty e^{iz\xi} \psi(\xi)\, d\xi$$

for $y \geq 0$, where $\psi(\xi) \in L^2(0, \infty)$. This equality can be rewritten in the form

$$\frac{1}{\varepsilon} \int_z^{z+\varepsilon} g(\zeta)\, d\zeta = \frac{1}{\sqrt{2\pi}} \int_{-A'}^\infty e^{itz} \psi(t + A')\, dt.$$

Setting $z = x$ and comparing the result with (7), we conclude that $\varphi(t) = 0$ for $t < -A'$. But since $A' > A$ is arbitrary, $\varphi(t) = 0$ for $t < -A$, and (7) takes the form

$$\frac{1}{\varepsilon} \int_x^{x+\varepsilon} g(\xi)\, d\xi = \frac{1}{\sqrt{2\pi}} \int_{-A}^\infty \varphi(t) \frac{e^{i\varepsilon t} - 1}{i\varepsilon t} e^{itx}\, dt. \tag{10}$$

It can be proved similarly that $\varphi(t) = 0$ for $t > A$. Hence, (10) can be written in the form

$$\frac{1}{\varepsilon} \int_x^{x+\varepsilon} g(\xi)\, d\xi = \frac{1}{\sqrt{2\pi}} \int_{-A}^A \varphi(t) \frac{e^{i\varepsilon t} - 1}{i\varepsilon t} e^{itx}\, dt.$$

To conclude the proof it now remains to pass to the limit as $\varepsilon \searrow 0$.

**6. Problems.** 1. Prove that if $F(z) \in H_2^+$, then

$$\lim_{y \to \infty} F(x + iy) = 0$$

uniformly with respect to $x$ $(-\infty < x < \infty)$ and

$$\lim_{\pm x \to \infty} F(x + iy) = 0$$

uniformly with respect to $y \geq y_0 > 0$.

*Hint.* Use Theorem 1 for proving the first assertion and the representation

$$F(z) = \frac{1}{2\pi i} \int_{-\infty}^\infty \frac{g(\xi)\, d\xi}{\xi - z}$$

for proving the second.

2. If $F(z)$ is regular for $y \geq 0$ and $|F(z)| < Ae^{-\delta y}$, where $A$ and $\delta$ are positive constants, and if $\int_{-\infty}^\infty |F(x)|^2\, dx < \infty$, then $F(z) \in H_2^+$ (prove this).

*Hint.* Use the Cauchy formula for the half-disk $|a| < R$, $\operatorname{Im} z > 0$, with $R$ large, and then use Theorem 3.

3. Let $\Phi(s)$ $(s = \sigma + i\tau)$ be a function regular at infinity, and suppose that the expansion

$$\Phi(s) = \sum_{n=0}^{\infty} \frac{a_n}{s^{n+1}}$$

is valid in the domain $|s| > c$ for some $c > 0$. Prove that in this case

$$\varphi(z) = \sum_{n=1}^{\infty} \frac{a_n}{n!} z^n$$

is an entire function of exponential type $\leq c$, and for $\sigma > c$ the representation

$$\Phi(s) = \int_0^{\infty} e^{-st} \varphi(t)\, dt$$

is valid, so that $\Phi(s)$ is the Laplace transform of $\varphi(t)$.

*Hint.* Write $\Phi(s)$ in the form

$$\Phi(s) = \sum_{n=0}^{\infty} \frac{a_n}{n!} \int_0^{\infty} e^{-sx} x^n\, dx.$$

CHAPTER 12

# Boundary Properties of Functions Analytic in the Upper Half-plane, and the Hilbert Transform

**1.** In this chapter we again consider analytic functions in the upper half-plane. These functions will now be given by integrals of Cauchy type. Suppose that $f(t)$ $(-\infty < t < \infty)$ is measurable and satisfies the condition

$$\int_{-\infty}^{\infty} \frac{|f(t)|}{1+|t|}\,dt < \infty. \tag{1}$$

We underscore that $f$ is not assumed to be real, except when this is stipulated explicitly. An integral of Cauchy type can be written in the form

$$\Phi(z) = \frac{1}{\pi i}\int_{-\infty}^{\infty} \frac{f(t)}{t-z}\,dt \qquad (\operatorname{Im} z > 0).$$

Since

$$\frac{1}{i}\cdot\frac{1}{t-z} = \frac{y}{(x-t)^2+y^2} + i\frac{x-t}{(x-t)^2+y^2},$$

it follows that $\Phi(z) = u(x, y) + iv(x, y)$, where

$$u(x, y) = \frac{1}{\pi}\int_{-\infty}^{\infty} \frac{f(t)y\,dt}{(x-t)^2+y^2},$$

$$v(x, y) = \frac{1}{\pi}\int_{-\infty}^{\infty} \frac{f(t)(x-t)\,dt}{(x-t)^2+y^2}.$$

Both $u(x, y)$ and $v(x, y)$ are harmonic in the upper half-plane, and they form what is commonly called a *conjugate harmonic pair*. The function $u(x, y)$ is a Poisson integral in the upper half-plane, and satisfies the relation $\lim_{y\searrow 0} u(x, y) = f(x)$ for almost all $x$ $(-\infty < x < \infty)$, by the Bochner representation theorem.

Our first problem is to investigate the behavior of $v(x, y)$ as $y \searrow 0$. This question was initially studied more than 50 years ago by Plessner.

THEOREM 1. *Under the single condition* (1) *the limit* $\lim_{y\searrow 0} v(x, y)$ *exists and is finite for almost all* $x$ $(-\infty < x < \infty)$.

PROOF. Here we first split $f(t)$ into its real and imaginary parts, $f(t) = f_1(t) + if_2(t)$, and then introduce the positive and negative parts of each of the functions $f_1(t)$ and $f_2(t)$. We get the representation

$$f(t) = f_1^+(t) + f_1^-(t) + i\{f_2^+(t) + f_2^-(t)\},$$

where each of the functions $f_k^\pm(t)$ is real, does not change sign on the axis, and satisfies condition (1). Therefore, it suffices to prove Theorem 1 for a function $f(t) \geq 0$. But in this case $u(x, y) \geq 0$ everywhere in the half-plane $y > 0$, and thus the function $\Psi(z) = \exp\{-\Phi(z)\}$ which is analytic in the domain $y > 0$, is bounded in this half-plane:

$$|\Psi(z)| = e^{-u(x,y)} \leq 1.$$

Hence, the function

$$F(z) = \frac{\Psi(z)}{z+i}$$

belongs to the class $H_2^+$, and hence(30) the limit

$$F(x) = F(x+i0) = \lim_{y\searrow 0} \frac{\Psi(x+iy)}{x+iy+i}$$

exists and is finite for almost all $x \in (-\infty, \infty)$. The same is true for the limit $\lim_{y\searrow 0} \Psi(x+iy)$, i.e., it exists and is finite a.e. Moreover, this last limit is a.e. nonzero, since otherwise the quantity

$$f(x) = \lim_{y\searrow 0} u(x, y) = -\lim_{y\searrow 0} \ln|\Psi(x+iy)|$$

would not be a.e. finite. It follows that the limit

$$\lim_{y\searrow 0} \Phi(x+iy) = -\lim_{y\searrow 0} \ln\{\Psi(x+iy)\}$$

exists and is finite a.e., and this gives us Theorem 1.

**2.** If the function $f$ satisfies (1), then for any $\varepsilon > 0$ the integral

$$\tilde{f}_\varepsilon(x) = \frac{1}{\pi}\int_{|t-x|>\varepsilon} \frac{f(t)}{x-t}\,dt = \frac{1}{\pi}\int_{-\infty}^{x-\varepsilon} + \frac{1}{\pi}\int_{x+\varepsilon}^{\infty} \frac{f(t)\,dt}{x-t}$$

exists and is finite. In analysis courses the limit of this expression as $\varepsilon \searrow 0$, if this limit exists and is finite, is called the *Cauchy principal value* of the integral and denoted by

$$\text{p.v.}\frac{1}{\pi}\int_{-\infty}^{\infty} \frac{f(t)}{x-t}\,dt.$$

We can now proceed to the proposition answering the question posed above.

(30) See the corollary to Theorem 2 in Chapter 11.

THEOREM 2 (PLESSNER). *Under the single condition* (1),

$$\lim_{y\searrow 0}\{v(x,y)-\tilde{f}_y(x)\}=0$$

*at all Lebesgue points of the function* $f$.

PROOF. From the definitions of $v(x,y)$ and $\tilde{f}_y(x)$ it follows that

$$\begin{aligned}v(x,y)-\tilde{f}_y(x)=&\frac{1}{\pi}\int_{|t-x|<y} f(t)\frac{(x-t)\,dt}{(x-t)^2+y^2}\\&+\frac{1}{\pi}\int_{|t-x|>y} f(t)\left[\frac{x-t}{(x-t)^2+y^2}-\frac{1}{x-t}\right]dt.\end{aligned}$$

We now introduce the kernel

$$\tilde{k}(z)=\begin{cases}\dfrac{1}{\pi}\cdot\dfrac{z}{z^2+1} & (|z|\le 1),\\ \dfrac{1}{\pi}\cdot\left\{\dfrac{z}{z^2+1}-\dfrac{1}{z}\right\} & (|z|>1).\end{cases}$$

With the help of this kernel we can rewrite the above relation in the form

$$v(x,y)-\tilde{f}_y(x)=\int_{-\infty}^{\infty} f(t)\frac{1}{y}\tilde{k}\left(\frac{x-t}{y}\right)dt.$$

The kernel $\tilde{k}(z)$ satisfies the condition

$$(1+|z|)^{1+\lambda}|\tilde{k}(z)|\le M<\infty$$

even for $\lambda=2$, but the condition $\int_{-\infty}^{\infty}\tilde{k}(z)\,dz\ne 0$ in the Bochner representation theorem is not satisfied. However, by using the technique in Chapter 1, §4, it can be shown([31]) that at all Lebesgue points of $f$

$$\lim_{y\searrow 0}\int_{-\infty}^{\infty} f(t)\frac{1}{y}\tilde{k}\left(\frac{x-t}{y}\right)dt=0.$$

We leave the verification of this assertion to the reader.

It is stressed that the following general fact, also established by Plessner, also follows from Theorems 1 and 2.

THEOREM 3. *If* $f$ *satisfies condition* (1), *then the principal value of the integral*

$$\tilde{f}(x)=\text{p.v.}\frac{1}{\pi}\int_{-\infty}^{\infty}\frac{f(t)\,dt}{x-t} \tag{2}$$

*exists for almost all* $x$.

---

([31])See Problem 4 in Chapter 1.

**3.** If $f$ belongs to one of the classes $L^p(-\infty, \infty)$ $(p \geq 1)$ then it certainly satisfies condition (1). The question arises as to whether $\tilde{f}$ belongs to the same class. A simple example shows that this can fail to be true for $p = 1$.

Let $-\infty < a < b < \infty$, and let

$$f(t) = \begin{cases} 1 & \text{for } t \in (a, b), \\ 0 & \text{for } t \notin (a, b). \end{cases}$$

By definition, for any $x$ except $x = a$ and $x = b$

$$\tilde{f}(x) = \frac{1}{\pi} \ln \left| \frac{x-a}{x-b} \right| = \frac{1}{\pi} \ln \left| 1 + \frac{b-a}{x-b} \right|.$$

Therefore

$$|\tilde{f}(x)| \sim \frac{b-a}{\pi |x|} \qquad (|x| \to \infty),$$

and hence $\tilde{f} \notin L^1(-\infty, \infty)$.

Matters are different for $p > 1$. Namely, we have the following remarkable

THEOREM (M. RIESZ). *If $f \in L^p(-\infty, \infty)$ for $1 < p < \infty$, then $\tilde{f} \in L^p(-\infty, \infty)$, $\|\tilde{f}\|_p \leq A_p \|f\|_p$, and*

$$f(x) = \text{p.v.} \frac{1}{\pi} \int_{-\infty}^{\infty} \frac{\tilde{f}(t)}{t-x} \, dt$$

*almost everywhere.*

Thus, for any finite $p > 1$ formula (2) determines a continuous linear operator mapping $L^p(-\infty, \infty)$ into itself. This operator is denoted by $\mathfrak{H}$ and called the *Hilbert transform.*([32]) The Riesz theorem implies also that $\mathfrak{H}$ is invertible and $\mathfrak{H}^{-1} = -\mathfrak{H}$. The Riesz theorem was proved in 1924, but the exact value of the constant $A_p$ (i.e., the norm of $\mathfrak{H}$) was found only in 1972.([33]) It is

$$A_p = \begin{cases} \tan(\pi/2p) & (1 < p \leq 2), \\ \cot(\pi/2p) & (2 \leq p < \infty). \end{cases}$$

Consideration of the Hilbert transform for arbitrary $p$ goes beyond the framework of this book, and we confine ourselves to the case $p = 2$, which is closely connected with the theory of the Fourier transform. In this case we prove that *the Hilbert transform is an integral operator on $L^2(-\infty, \infty)$, and $\mathfrak{H}^{-1} = -\mathfrak{H}$.* This yields the assertion of the Riesz theorem for $p = 2$ with the constant $A_2 = 1$.

---

([32])It would be more correct to call it the Riesz-Hilbert transform, since Hilbert essentially considered a somewhat different operator, and that only for $p = 2$.

([33])See S. K. Pichorides, *On the best values of the constants in the theorems of M. Riesz, Zygmund and Kolmogorov*, Studia Math. **44** (1972), 165–179.

**4.** Accordingly, consider the case $p = 2$. Each function $f \in L^2(-\infty, \infty)$ can be represented as the inverse Fourier transform of some function $h \in L^2(-\infty, \infty)$,

$$f(t) = \frac{1}{\sqrt{2\pi}} \int_{-\infty}^{\infty} h(\xi) e^{it\xi}\, d\xi, \tag{3}$$

and $f$ runs through the whole space $L^2(-\infty, \infty)$ as $h$ does. We now introduce another function $g \in L^2(-\infty, \infty)$, defined by

$$g(t) = \frac{1}{\sqrt{2\pi}} \int_{-\infty}^{\infty} h(\xi) \frac{\operatorname{sgn} \xi}{i} e^{it\xi}\, d\xi. \tag{3'}$$

Formulas (3) and (3′) can be regarded as a parametric representation of some linear operator $U$ carrying $f$ into $g$. Since $\|f\|_2 = \|h\|_2 = \|g\|_2$, $U$ is unitary. Note that if $f$ or $g$ is real, then so is the other. Indeed, if, for example, $f$ is real, then its Fourier transform $h$ is an Hermitian function; that is, $h(-t) = \overline{h(t)}$. But in this case the function $h_1(t) = h(t)(\operatorname{sgn} t)/i$ is also Hermitian, and hence its inverse Fourier transform $g(t)$ is a real function.

We prove that

$$g(x) = \tilde{f}(x) \tag{4}$$

almost everywhere, where $\tilde{f}(x)$ is defined by (2), and, moreover,

$$f(x) = -\tilde{g}(x) \tag{4'}$$

almost everywhere. Formula (4) shows, first, that for $f \in L^2(-\infty, \infty)$ the function $\tilde{f}(x)$ also belongs to $L^2(-\infty, \infty)$, and second, the operator $U$ actually coincides with the Hilbert transform, i.e., $U = \mathfrak{H}$. Then it follows from (4′) that $U^{-1} = -\mathfrak{H}$. The properties of $\mathfrak{H}$ on $L^2(-\infty, \infty)$ formulated in §3 thus follow from (4) and (4′).

To prove (4) and (4′) it clearly suffices to assume that $f \in L^2(-\infty, \infty)$ is real, and then $g \in L^2(-\infty, \infty)$ has the same property. As in §1, we introduce the function

$$\Phi(z) = \frac{1}{\pi i} \int_{-\infty}^{\infty} \frac{f(t)\, dt}{t - z} = u(x, y) + iv(x, y) \qquad (y = \operatorname{Im} z > 0),$$

where

$$\begin{aligned} u(x, y) &= \frac{1}{\pi} \int_{-\infty}^{\infty} \frac{f(t) y\, dt}{(x - t)^2 + y^2}, \\ v(x, y) &= \frac{1}{\pi} \int_{-\infty}^{\infty} \frac{(x - t) f(t)\, dt}{(x - t)^2 + y^2}. \end{aligned} \tag{5}$$

Let us apply the Parseval equality to the integral representing $\Phi(z)$, just as in §2 of Chapter 11, obtaining

$$\Phi(z) = \sqrt{\frac{2}{\pi}} \int_0^{\infty} f(t) e^{itz}\, dt = \sqrt{\frac{2}{\pi}} \int_0^{\infty} h(t) e^{itz}\, dt. \tag{6}$$

We now introduce the new function

$$\Psi(z) = \frac{1}{\pi i}\int_{-\infty}^{\infty}\frac{g(t)\,dt}{t-z} \qquad (y = \operatorname{Im} z > 0)$$

and find in this way the representation

$$\Psi(z) = \sqrt{\frac{2}{\pi}}\int_0^{\infty}\hat{g}(t)e^{itz}\,dt = \sqrt{\frac{2}{\pi}}\int_0^{\infty}h(t)\frac{\operatorname{sgn} t}{i}e^{itz}\,dt. \tag{7}$$

By (6) and (7), $\Psi(z) = (1/i)\Phi(z) = v(x, y) - iu(x, y)$. We write the formulas analogous to (5) for the real and imaginary parts of $\Psi(z)$:

$$\begin{aligned} v(x, y) &= \frac{1}{\pi}\int_{-\infty}^{\infty}\frac{g(t)y\,dt}{(x-t)^2+y^2}, \\ -u(x, y) &= \frac{1}{\pi}\int_{-\infty}^{\infty}\frac{(x-t)g(t)\,dt}{(x-t)^2+y^2}. \end{aligned} \tag{5'}$$

Comparing now the second of the formulas (5) with the first of the formulas (5′), letting $y$ go to 0, and using Theorem 2, we get (4). Comparison of the second pair of formulas in (5) and (5′) yields (4′).

5. PROBLEMS. 1. Verify that the Hilbert transform of the function $f(t) = 1/\pi(t^2+1)$ has the form $\tilde{f}(t) = t/\pi(t^2+1)$. This shows once again that for $f \in L^1(-\infty,\infty)$, $\tilde{f}$ can fail to belong to $L^1(-\infty,\infty)$.

2. The cosine transform and the sine transform of a function $a(t) \in L^2(0,\infty)$ are defined by

$$\mathfrak{F}_c a(t) = \sqrt{\frac{2}{\pi}}\,\underset{A\to\infty}{\text{l. i. m.}}\int^{A} a(\xi)\cos t\xi\,d\xi,$$

$$\mathfrak{F}_s a(t) = \sqrt{\frac{2}{\pi}}\,\underset{A\to\infty}{\text{l. i. m.}}\int_0^{A} a(\xi)\sin t\xi\,d\xi.$$

The operator $\mathfrak{F}_c$ (respectively, $\mathfrak{F}_s$) maps $L^2(0,\infty)$ in one-to-one fashion onto the subspace of even (respectively, odd) functions in $L^2(-\infty,\infty)$. Show that if $f = a + b$ is the decomposition of an arbitrary function in $L^2(-\infty,\infty)$ into the sum of an even and an odd function, then $\mathfrak{H}f = -\mathfrak{F}_c\mathfrak{F}_s^{-1}b + \mathfrak{F}_s\mathfrak{F}_c^{-1}a$.

3. Prove that the sine transform of $f(t) = (\sin t)/t$ is

$$\mathfrak{F}_s f(t) = \frac{1}{\sqrt{2\pi}}\ln\left|\frac{1+t}{1-t}\right|.$$

4. Prove that

$$\frac{1}{2\pi}\int_0^{\infty}\ln\left|\frac{\alpha+x}{\alpha-x}\right|\ln\left|\frac{\beta+x}{\beta-x}\right|dx = \frac{\pi}{2}\min(\alpha,\beta) \qquad (\alpha,\beta>0).$$

CHAPTER 13

# The Poisson Summation Formula and Some of Its Applications

**1.** This chapter is devoted to the so-called *Poisson summation formula*. One of the simplest variants of this formula has the form

$$\sqrt{\alpha}\cdot\sum_{k=-\infty}^{\infty} f(k\alpha)=\sqrt{\beta}\cdot\sum_{k=-\infty}^{\infty}\hat{f}(k\beta), \tag{1}$$

where $\alpha\beta = 2\pi$ and $\alpha > 0$. For (1) to hold it suffices that $f(x)$ have bounded variation on the whole axis and belong to $L^1(-\infty,\infty)$. It is assumed that at points of discontinuity of $f$

$$f(x)=\frac{f(x+0)+f(x-0)}{2}.$$

A very useful special case of (1) is that in which $f(x)=e^{-x^2/2}$ and, hence, $\hat{f}(x)=e^{-x^2/2}$. In this case (1) takes the form

$$\sqrt{\alpha}\left(\frac{1}{2}+\sum_{k=1}^{\infty}e^{-\alpha^2k^2/2}\right)=\sqrt{\beta}\left(\frac{1}{2}+\sum_{k=1}^{\infty}e^{-\beta^2k^2/2}\right). \tag{2}$$

Incidentally, this implies the relation

$$\lim_{\alpha\searrow 0}\alpha\left(\frac{1}{2}+\sum_{k=1}^{\infty}e^{-\alpha^2k^2/2}\right)=\sqrt{\frac{\pi}{2}}.$$

Formula (2) and its generalizations play a fairly essential role in the theory of elliptic theta-functions. If it is assumed that $f(x)$ is even and equal to zero for $|x|>A$, then (1) takes the form

$$\frac{1}{2}+\sum_{k=1}^{n}f(k\alpha)=\frac{1}{2\alpha}\int_0^A f(x)\,dx+\frac{1}{\alpha}\sum_{k=1}^{\infty}\int_0^A f(x)\cos\frac{2k\pi x}{\alpha}\,dx, \tag{1'}$$

and $n=[A/\alpha]$. This formula should be compared with the Euler-Maclaurin formula in the so-called calculus of finite differences. The second term on the right-hand side of (1′) can be regarded as the error committed in replacing the finite sum (the left-hand side) by the integral (the first term on the right-hand side). In our subsequent considerations we shall be interested first and foremost in the formal aspect of the generalization of (1) to functions on $E_n$. We conclude the chapter with an interesting application.

**2.** It is convenient for us to employ somewhat modified definitions and notation for the coefficients and the Fourier transform. We begin by regarding the periodic functions with period 1 instead of $2\pi$. If $F(x) \in L^1(T_n)$, where

$$T_n = \{x \in E_n | 0 \leq x_j \leq 1;\ j = 1, 2, \dots, n\},$$

then the Fourier coefficients are determined by

$$F^*(m) = \int_{T_n} F(x) e^{-2\pi i(x,m)}\, dx,$$

where $m$ is an integer multi-index $\{m_1, \dots, m_n\}$, and the Fourier series of $F(x)$ has the form

$$F(x) \sim \sum_{(m)} F^*(m) e^{2\pi i(x,m)}.$$

Further, if $f(x) \in L^1(E_n)$, then the Fourier transform is

$$f^*(t) = \int_{E_n} f(x) e^{-2\pi i(x,t)}\, dx.$$

Thus, the new notation is connected with the old by the formula

$$f^*(t) = (\sqrt{2\pi})^n \hat{f}(2\pi t).$$

The corresponding inversion formula has the form

$$f(x) = \int_{E_n} f^*(t) e^{2\pi i(x,t)}\, dt,$$

and the Parseval equality for $f \in L^2(E_n)$ looks the same as before:

$$\int_{E_n} |f^*(t)|^2\, dt = \int_{E_n} |f(x)|^2\, dx.$$

For convenience we set

$$T_n^{(m)} = \{x \in E_n | m_j \leq x_j \leq m_j + 1;\ j = 1, 2, \dots, n\}$$

for any integer multi-index $m$. Thus, $T_n = T_n^{(0)}$.

If $f(x) \in L^1(E_n)$, then $\sum_{(m)} f(x+m)$ converges in the $L^1(T_n)$-metric to some function $F(x) \in L^1(T_n)$:

$$F(x) = \sum_{(m)} f(x+m). \tag{3}$$

Indeed,

$$\int_{T_n} dx \cdot \sum_{|m|<N} |f(x+m)| = \sum_{|m|<N} \int_{T_n^{(0)}} |f(x+m)|\,dx$$
$$= \sum_{|m|<N} \int_{T_0^{(m)}} |f(x)|\,dx \le \int_{E_n} |f(x)|\,dx.$$

For any integer multi-index $k$ we can write

$$F(x)e^{-2\pi i(x,k)} = \sum_{(m)} f(x+m)e^{-2\pi i(x+m,k)}$$

together with (3). Since this series converges in the $L^1(T_n)$-metric,

$$F^*(k) = \int_{T_n} F(x)e^{-2\pi i(x,k)}\,dx = \int_{T_n} \sum_{(m)} f(x+m)e^{-2\pi i(x+m,k)}\,dx$$
$$= \sum_{(m)} \int_{T_n} f(x+m)e^{-2\pi i(x+m,k)}\,dx = \sum_{(m)} \int_{T_n^{(m)}} f(x)e^{-2\pi i(x,k)}\,dx$$
$$= \int_{E_n} f(x)e^{-2\pi i(x,k)}\,dx = f^*(k).$$

Therefore, the Fourier series expansion of $F(x)$ can be written in the form

$$F(x) = \sum_{(m)} f(x+m) = \sum_{(k)} f^*(k)e^{2\pi i(x,k)}, \tag{4}$$

where we have replaced the sign $\sim$ by $=$, without investigating conditions under which the series converges.

We generalize this result somewhat, introducing the parameter $y \in E_n$ and replacing $f(x)$ in (4) by $f(x)e^{-2\pi i(x,y)}$, and hence $f^*(t)$ by $f^*(t+y)$:

$$\sum_{(m)} f(x+m)e^{-2\pi i(x+m,y)} = \sum_{(k)} f^*(y+k)e^{2\pi i(x,k)}.$$

Multiplying both sides by $e^{\pi i(x,y)}$, we get the final formula

$$\sum_{(m)} f(x+m)e^{-2\pi i(m+x/2,y)} = \sum_{(k)} f^*(y+k)e^{2\pi i(x,k+y/2)}, \tag{5}$$

which has a symmetric form. For $y = 0$ this becomes (4).

**3.** We proceed to a further generalization of (4). With this goal we take a nonsingular $n \times n$ matrix $A$ and let $x' = Ax$. Denote by $A'$ the transpose matrix, and by $B$ the matrix inverse to $A'$. We now carry out some obvious transformations:

$$f^*(y) = \int_{E_n} f(x')e^{-2\pi i(y,x')}\,dx' = |\det A| \cdot \int_{E_n} f(Ax)e^{-2\pi i(y,Ax)}\,dx$$
$$= |\det A| \cdot \int_{E_n} f(Ax)e^{-2\pi i(A'y,x)}\,dx.$$

Hence

$$f^*(By) = |\det A| \cdot \int_{E_n} f(Ax)e^{-2\pi i(y,x)}\,dx,$$

or (since $\det A \times \det B = 1$)

$$\sqrt{|\det B|} \cdot f^*(By) = \sqrt{|\det A|} \cdot \int_{E_n} f(Ax)e^{-2\pi i(y,x)}\,dx.$$

In (5) we can replace $f(x)$ by $\sqrt{|\det A|}f(Ax)$, and then $f^*(y)$ must be replaced by $\sqrt{|\det B|}f^*(By)$. Then (5) is replaced by the more general formula

$$\begin{aligned}\sqrt{|\det A|}\sum_{(m)} f(Ax + Am)e^{-2\pi i(m+x/2,y)}\\ = \sqrt{|\det B|}\sum_{(k)} f^*(By + Bk)e^{2\pi i(x,k+y/2)}.\end{aligned} \tag{6}$$

**4.** Formula (6) has various applications in number theory. We consider one of them, associated with the name of Siegel. We write (6) with $x = y = 0$, assuming that $\det A > 0$ and denoting the components of the vectors $Ax$ and $By$ by $a_p(x)$ and $b_q(y)$. The function $f$ is taken to be

$$f(x) = \begin{cases} \prod_{k=1}^{n}(t_k - |x_k|) & \text{if } |x_k| < t_k \text{ for each } k = 1, \dots, n, \\ 0 & \text{otherwise.}\end{cases}$$

In this case $f^*$ is the Fejér kernel

$$f^*(y) = \prod_{k=1}^{n} \frac{\sin^2(\pi t_k y_k)}{\pi^2 y_k^2}.$$

Then (6) takes the form

$$\det A \cdot \sum_{(m)}\prod_{k=1}^{n}(t_k - |q_k(m)|)^+ = \sum_{(m)}\prod_{k=1}^{n}\frac{\sin^2[\pi t_k b_k(m)]}{\pi^2 b_k^2(m)}. \tag{7}$$

LEMMA. *If for each nontrivial $n$-tuple $m$ there is an $i$ such that $|a_i(m)| \geq t_i$, then $t_1 \cdots t_n \leq \det A$.*

PROOF. Under the condition of the lemma the left-hand side of (7) consists of only a single term corresponding to the trivial $n$-tuple, and hence the left-hand side is equal to $\det A \cdot t_1 \cdots t_n$, while the right-hand side of (7), which consists only of nonnegative terms, is not less than the term corresponding to the trivial $n$-tuple, and is thus equal to $(t_1 \cdots t_n)^2$. Therefore, $\det A \cdot t_1 \cdots t_n \geq (t_1 \cdots t_n)^2$, which implies that $t_1 \cdots t_n \leq \det A$. The following result is a consequence of the lemma.

MINKOWSKI THEOREM. *If the linear forms with real coefficients*

$$a_p(x) = \sum_{q=1}^{n} a_{pq}x_q \quad (p = 1, \dots, n)$$

*have determinant* $\det A = 1$, *then there is at least one nontrivial integer n-tuple* $m = \{m_1, \dots, m_n\}$ *such that* $|a_1(m)| \leq 1$ *and* $|a_i(m)| < 1$ $(i = 2, 3, \dots, n)$.

PROOF. Let $t_1 = 1 + \varepsilon$ and $t_2 = t_3 = \cdots = t_n = 1$, where $\varepsilon > 0$. Since $t_1 \cdots t_n > \det A$, the lemma gives us the existence of at least one nontrivial $n$-tuple such that

$$|a_1(m)| < 1 + \varepsilon, \qquad |a_j(m)| < 1 \quad (j = 2, 3, \dots, n). \tag{8}$$

Take $\varepsilon < 1$. Then the integer point corresponding to this $n$-tuple lies in the domain $Q \subset E_n$ bounded by the planes $a_1(x) = \pm 2$, and $a_i(x) = \pm 1$ $(i = 2, \dots, n)$. This domain contains only finitely many integer points. Since the indicated $n$-tuple exists for each $\varepsilon > 0$, by taking an infinite sequence $\varepsilon_i \to 0$ we conclude that at least one nontrivial $n$-tuple satisfying (8) is encountered infinitely many times. For this nontrivial $n$-tuple

$$|a_1(m)| < 1 + \varepsilon_i, \qquad |a_j(m)| < 1 \quad (j = 2, 3, \dots, n).$$

Hence, taking the limit as $\varepsilon_i \to 0$, we arrive at the result to be proved.

We remark that any of the numbers $2, \dots, n$ could have been taken as the distinguished index instead of 1. However, it cannot be required that the sign $<$ appears everywhere. Here is an example:

$$a_1(x) = 2x_1 + x_2, \qquad a_2(x) = x_1 + x_2,$$

so that $\det A = 1$. For any integer pair $x = \{x_1, x_2\}$ the forms $a_1(x)$ and $a_2(x)$ are integers. If they are less than 1 in modulus, then both are hence equal to zero.

A simpler geometric proof of the Minkowski theorem, along with applications of it, can be found in books on number theory.([34]) However, Siegel's proof presented here is also very interesting.

---

([34])See, for example, Z. I. Borevich and I. R. Shafarevich, *Number theory*, 2nd rev. ed., "Nauka", Moscow, 1972, Chapter 2; English transl., Academic Press, 1966.

CHAPTER 14

# Applications of the Laplace and Fourier Transforms to the Solution of Boundary Value Problems in Mathematical Physics

We show by some examples how the Laplace and Fourier transforms are used in the solution of boundary value problems in mathematical physics. Only the analytic aspects will be considered; the physical content of the problems will not be touched upon.

EXAMPLE 1. Find the solution of the heat equation

$$\frac{\partial u}{\partial t} = \frac{\partial^2 u}{\partial x^2} \qquad (x > 0, t > 0)$$

under the boundary conditions([35])

$$u(0, t) = 0 \quad (t > 0), \qquad u(x, 0) = 1 \quad (x > 0).$$

The solution is sought in the class of functions such that both $u(x, t)$ and its first and second partial derivatives with respect to $x$ remain bounded as $t \to \infty$ for any $x > 0$. We write the Laplace transform of the unknown solution

$$U(x, s) = \int_0^\infty u(x, t)e^{-st}\,dt \qquad (s = \sigma + i\tau)$$

and pose the problem of determining $U(x, s)$. In view of our assumption about the boundedness of the functions under consideration as $t \to \infty$ we can write the equality

$$\begin{aligned}\frac{\partial^2 U(x, s)}{\partial x^2} &= \int_0^\infty \frac{\partial^2 u(x, t)}{\partial x^2} e^{-st}\,dt = \int_0^\infty \frac{\partial u(x, t)}{\partial t} e^{-st}\,dt \\ &= e^{-st}u(x, t)|_{t=0}^{\infty} + s\int_0^\infty u(x, t)e^{-st}\,dt \\ &= -1 + sU(x, s) \qquad (x > 0)\end{aligned}$$

([35])Here

$$u(0, t) = \lim_{x \searrow 0} u(x, t), \quad u(x, 0) = \lim_{t \searrow 0} u(t, x),$$

and the analogous notation is used in what follows.

for $\sigma > 0$. This ordinary second-order differential equation with respect to $x$ has the general integral

$$U(x,s) = \frac{1}{s} + Ae^{-x\sqrt{s}} + Be^{x\sqrt{s}}.$$

In the $s$-plane with a cut along the semi-axis $\tau = 0$, $\sigma < 0$ we choose a continuous branch of $\sqrt{s}$ such that $\operatorname{Re}\sqrt{s} > 0$. Observing that, since $u(x,t)$ is bounded as $t \to \infty$, $U(x,s)$ must remain bounded as $\sigma \to \infty$, we conclude that $B = 0$. On the other hand,

$$\frac{1}{s} + A = \lim_{x \searrow 0} U(x,s) = \lim_{x\searrow 0} \int_0^\infty u(x,t)e^{-st}\,dt = 0,$$

so that $A = -1/s$, and hence

$$U(x,s) = \frac{1 - e^{-x\sqrt{s}}}{s}.$$

It is now necessary to use the inversion formula (5) in Chapter 11 for the Laplace transform. If the limit

$$\lim_{T\to\infty} \frac{1}{2\pi i}\int_{\sigma - iT}^{\sigma + iT} U(x,s)e^{st}\,ds \qquad (\sigma > 0) \tag{$*$}$$

exists for all $x, t > 0$, then for each $x > 0$ it coincides almost everywhere on the semi-axis $t > 0$ with the corresponding limit in the mean, which is equal to $u(x,t)$ on the basis of the formula just cited. If, moreover, the limit $(*)$ depends continuously on $x$ and $t$, then it is equal to $u(x,t)$ everywhere. We therefore need to investigate the integral

$$\int_{\sigma - iT}^{\sigma + iT} \frac{1 - e^{-x\sqrt{s}}}{s} e^{st}\,ds$$

for large $T$, where $x$ and $t$ are fixed positive numbers. For this we use the theory of contour integrals, noting that the only singular point of the integrand is the point $s = 0$. We deform our path of integration (a segment parallel to the imaginary axis) first into a path consisting of the two symmetric segments $\varepsilon \le |\tau| \le T$ of the imaginary axis and the semicircle $|s| = \varepsilon$, $\operatorname{Re} s \ge 0$, and then into a path consisting of the two segments going along the upper and lower sides of the cut on the negative semi-axis, and the circle $|s| = \delta < \varepsilon$. With this goal in mind, consider the closed contours pictured in Figures 1 and 2. Applying the Cauchy theorem to them successively and estimating the "superfluous" integrals, we get that

$$\begin{aligned}\int_{\sigma - iT}^{\sigma + iT} \frac{1 - e^{-x\sqrt{s}}}{s} e^{st}\,ds &= \int_{-iT}^{-i\varepsilon} + \int_{i\varepsilon}^{iT} \frac{1 - e^{-x\sqrt{s}}}{s} e^{st}\,ds + o(1)\\ &= \int_{-T}^{-\delta} \frac{1 - e^{-x(-ir)}}{s} e^{st}\,ds + \int_{-\delta}^{-T} \frac{1 - e^{-x(ir)}}{s} e^{st}\,ds + o(1)\end{aligned}$$

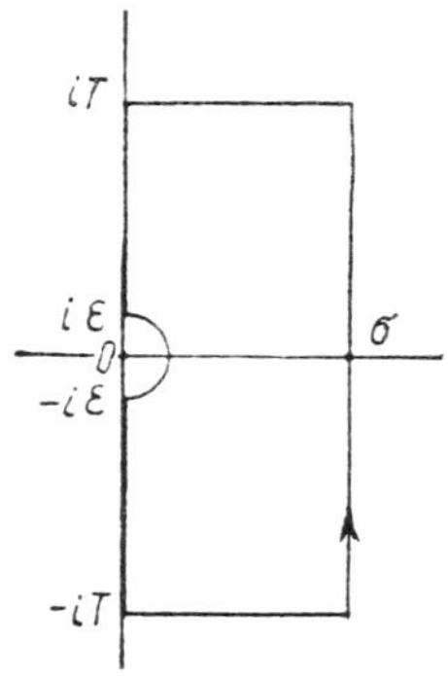


FIGURE 1

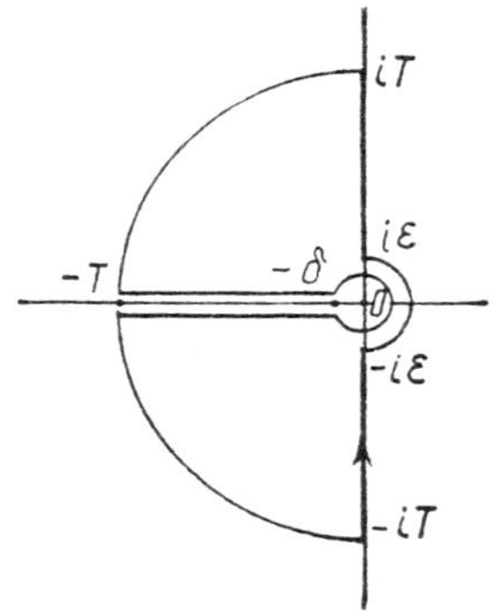


FIGURE 2

$(\delta, \varepsilon \to 0; T \to \infty)$, where $s = -r^2$, $r > 0$. Replacing $ds$ by $-2rdr$, we find that

$$\int_{\sigma-iT}^{\sigma+iT} \frac{1-e^{-x\sqrt{s}}}{s} e^{st}\, ds = 4i \int_{\sqrt{\delta}}^{\sqrt{T}} \frac{\sin xr}{r} e^{-tr^2}\, dr + o(1)$$

$(\delta \to 0; T \to \infty)$, from which we get, finally,

$$u(x,t) = \lim_{T\to\infty} \frac{1}{2\pi i} \int_{\sigma-iT}^{\sigma+iT} \frac{1-e^{-x\sqrt{s}}}{s} e^{st}\, ds = \frac{2}{\pi} \int_0^\infty \frac{\sin xr}{r} e^{-tr^2}\, dr.$$

Note that the solution can be given the form

$$u(x,t) = \frac{1}{\sqrt{\pi t}} \int_0^x e^{-y^2/4t}\, dy = \frac{2}{\sqrt{\pi}} \int_0^{x/2\sqrt{t}} e^{-\xi^2}\, d\xi.$$

EXAMPLE 2. Find a solution of the Laplace equation

$$\frac{\partial^2 u}{\partial x^2} + \frac{\partial^2 u}{\partial y^2} = 0 \qquad (x > 0, y > 0)$$

under the boundary conditions

$$u(x,0) = 0 \quad (x > 0), \qquad \frac{\partial u}{\partial x}(0,y) = \begin{cases} -q & (0 < y < b), \\ 0 & (y > b), \end{cases}$$

where $q = \text{const}$. Moreover, find the magnitude of the flow $q(x,0) = (\partial u/\partial y)(x,0)$ through the horizontal boundary of the domain.

The particular solution satisfying $u(x,0) = 0$ has the form $e^{-\lambda x} \sin \lambda y$, where $\lambda > 0$, since the solution is sought in the class of bounded functions. Therefore, it is possible to find $u(x,y)$ in the form

$$u(x,y) = \int_0^\infty A(\lambda) e^{-\lambda x} \sin \lambda y\, d\lambda.$$

By assumption,

$$-\frac{\partial u}{\partial x}(0, y) = \int_0^\infty A(\lambda)\lambda \sin \lambda y \, d\lambda = \begin{cases} 0 & (y > b), \\ q & (0 < y < b). \end{cases}$$

Therefore, by the inversion formula,

$$A(\lambda)\lambda = \frac{2}{\pi}\int_0^b q \sin \lambda y \, dy = \frac{2q(1-\cos b\lambda)}{\pi\lambda},$$

and hence

$$u(x, y) = \frac{2q}{\pi}\int_0^\infty \frac{1-\cos b\lambda}{\lambda^2} e^{-\lambda x} \sin \lambda y \, d\lambda,$$

and

$$q(x, 0) = \frac{2q}{\pi}\int_0^\infty \frac{1-\cos b\lambda}{\lambda} e^{-\lambda x} \, d\lambda = \frac{q}{\pi}\ln\left(1 + \frac{b^2}{x^2}\right).$$

EXAMPLE 3. Find the solution of the heat equation

$$\frac{\partial u}{\partial t} = \sum_{i=1}^n \frac{\partial^2 u}{\partial x_i^2} \qquad (x \in E_n; t > 0) \tag{1}$$

under the condition

$$u(x, 0) = f(x) \qquad (x \in E_n), \tag{2}$$

where $f(x)$ is a given continuous bounded function.

We show that the solution can be represented by the integral

$$u(x, t) = \frac{1}{(2\sqrt{\pi t})^n}\int_{E_n} f(y)e^{-|x-y|^2/4t} \, dy.$$

Indeed, the Weierstrass kernel(36)

$$\tilde{W}(z, t) = \frac{1}{(2\sqrt{\pi t})^n} e^{-|z|^2/4t}$$

satisfies the equation

$$\frac{\partial \tilde{W}}{\partial t} = \sum_{i=1}^n \frac{\partial^2 \tilde{W}}{\partial z_i^2}.$$

This implies that our function is a solution of (1). The boundary condition (2) follows from the Bochner representation theorem.

EXAMPLE 4. Find a solution of the heat equation

$$\frac{\partial u}{\partial t} = \frac{\partial^2 u}{\partial x^2} + \frac{\partial^2 u}{\partial \eta^2} \qquad (-\infty < x < \infty; \eta > 0; t > 0) \tag{3}$$

(36) $\tilde{W}(z, t) = W(z, 2t)$, where the kernel $W$ is defined by formula $(5_1)$ in Chapter 2.

under the conditions

$$u(x,\eta,0) = 0 \qquad (-\infty < x < \infty; \eta > 0), \tag{4}$$

$$u(x,0,t) = f(x) \qquad (-\infty < x < \infty; t > 0), \tag{5}$$

where $f(x)$ is a given continuous bounded function.

We show that the solution has the form

$$u(x,\eta,t) = \frac{1}{\pi}\int_{-\infty}^{\infty} f(y)\frac{\eta e^{-((x-y)^2+\eta^2)/4t}}{(x-y)^2+\eta^2}\,dy. \tag{6}$$

Since the kernel

$$K(z,\eta,t) = \frac{\eta\cdot e^{-(z^2+\eta^2)/4t}}{z^2+\eta^2} \qquad (-\infty < z < \infty; \eta > 0; t > 0)$$

satisfies the equation

$$\frac{\partial u}{\partial t} = \frac{\partial^2 u}{\partial z^2} + \frac{\partial^2 u}{\partial \eta^2},$$

the function (6) satisfies (3). It is clear from the estimate

$$|u(x,\eta,t)| \le \frac{1}{\pi}e^{-\eta^2/4t}\int_{-\infty}^{\infty}|f(y)|\frac{\eta}{(x-y)^2+\eta^2}\,dy$$

that this function satisfies (4). Finally, (6) can be represented in the form

$$\begin{aligned} u(x,\eta,t) =& \frac{1}{\pi}\int_{|y-x|<\delta} f(y)\frac{\eta\,dy}{(x-y)^2+\eta^2} \\ &+ \frac{1}{\pi}\int_{|y-x|>\delta} f(y)\frac{\eta}{(x-y)^2+\eta^2}e^{-((x-y)^2+\eta^2)/4t}\,dy \\ &+ \frac{1}{\pi}\int_{|y-x|<\delta} f(y)\frac{\eta}{(x-y)^2+\eta^2}[e^{-((x-y)^2+\eta^2)/4t}-1]\,dy \\ =& I_1 + I_2 + I_3. \end{aligned}$$

According to the theory of the Poisson integral, $I_1 \to f(x)$ and $I_2 \to 0$ as $\eta \searrow 0$ for fixed $\delta > 0$, the latter because

$$|I_2| \le \frac{1}{\pi}\int_{|y-x|>\delta}|f(y)|\frac{\eta}{(x-y)^2+\eta^2}\,dy.$$

For $I_3$ we have the estimate

$$|I_3| \le \frac{1}{\pi}\int_{|y-x|<\delta}|f(y)|\frac{\eta}{(x-y)^2+\eta^2}[1-e^{-\delta^2/4t}]\,dy < M(1-e^{-\delta^2/4t}),$$

where $M = \sup|f(y)|$, and the factor multiplying $M$ can be made arbitrarily small by suitably choosing $\delta$. Consequently, the function (6) satisfies the condition (5).

PROBLEMS.([37]) 1. Show that the solution of the nonhomogeneous heat equation

$$\frac{\partial u}{\partial t} = \frac{\partial^2 u}{\partial x^2} + f(x,t) \qquad (-\infty < x < \infty; t > 0)$$

under the condition $u(x,0) = 0$ $(-\infty < x < \infty)$ has the form

$$u(x,t) = \frac{1}{2\sqrt{\pi}} \int_0^t d\tau \int_{-\infty}^{\infty} f(y,t) e^{-(x-y)^2/4(t-\tau)} \frac{dy}{\sqrt{t-\tau}}.$$

2. Show that the solution of the heat equation

$$\frac{\partial u}{\partial t} = \frac{\partial^2 u}{\partial x^2} \qquad (x, t > 0)$$

under the conditions $u(0,t) = \varphi(t)$ $(t > 0)$, $u(x,0) = 0$ $(x > 0)$ has the form

$$u(x,t) = \frac{x}{2\sqrt{\pi}} \int_0^t \varphi(\tau) \frac{e^{-x^2/4(t-\tau)}}{(t-\tau)^{3/2}} d\tau.$$

3. Show that the solution of the heat equation

$$\frac{\partial u}{\partial t} = \frac{\partial^2 u}{\partial x^2} \qquad (x, t > 0)$$

under the conditions $u(0,t) = 0$ $(t > 0)$, $u(x,0) = f(x)$ $(x > 0)$ has the form

$$u(x,t) = \frac{1}{2\sqrt{\pi t}} \int_0^{\infty} f(y) \{e^{-(x-y)^2/4t} - e^{-(x+y)^2/4t}\} dy.$$

([37])The functions $f$ and $\varphi$ in the formulation of the boundary value problems are assumed for simplicity to be continuous and bounded.

CHAPTER 15

# Fourier Transforms of Increasing Functions. The Wiener-Hopf Technique

**1.** In this chapter we consider the classical generalization of the Fourier transform to the case of increasing functions. This generalization is connected with some propositions established in Chapter 11, and with the function class $H_2(\alpha, \beta)$ introduced there. We present (without proof) certain facts following in a simple way from the constructions in Chapter 11.

Let $f(t)$ $(-\infty < t < \infty)$ be a given function such that for some two real numbers $\alpha$ and $\beta$

$$\int_0^\infty |f(t) \cdot e^{-\alpha t}|^2 \, dt < \infty, \qquad \int_{-\infty}^0 |f(t) \cdot e^{-\beta t}|^2 \, dt < \infty. \tag{1}$$

We introduce two analytic functions of $z = x + iy$:[38]

$$F^+(z) = \int_0^\infty f(t) e^{itz} \, dt; \qquad F^-(z) = \int_{-\infty}^0 f(t) e^{itz} \, dt. \tag{2}$$

It is easy to see that $F^+(z) \in H_2(\alpha, \infty)$ and $F^-(z) \in H_2(-\infty, \beta)$.

**2.** In this case the half-planes $y > \alpha$ and $y < \beta$ have a common part when $\alpha < \beta$, namely, the strip $\alpha < y < \beta$, in which both functions $F^+(z)$ and $F^-(z)$ are regular. Hence in this strip we can consider the sum

$$F(z) \overset{\text{def}}{=} F^+(z) + F^-(z) = \int_{-\infty}^\infty f(t) e^{itz} \, dt. \tag{3}$$

This sum represents a function of class $H_2(\alpha, \beta)$. It follows from (3) that for any $\gamma \in (\alpha, \beta)$

$$F(x + i\gamma) = \int_{-\infty}^\infty f(t) e^{-\gamma t} e^{itx} \, dt.$$

---

([38]) The coefficient $1/\sqrt{2\pi}$ is not written here. Accordingly, the coefficient $1/2\pi$ appears below in (4) and analogous formulas.

Thus

$$f(t)e^{-\gamma t} = \frac{1}{2\pi}\int_{-\infty}^{\infty} F(x+i\gamma)e^{-itx}\,dx,$$

or

$$f(t) = \frac{1}{2\pi}\int_{-\infty}^{\infty} F(x+i\gamma)e^{-it(x+i\gamma)}\,dx,$$

and it is convenient to rewrite this formula (as already done in Chapter 11) in the form

$$f(t) = \frac{1}{2\pi}\int_{i\gamma-\infty}^{i\gamma+\infty} F(z)e^{-itz}\,dz. \tag{4}$$

This is the inversion formula for (3), where $\gamma$ is any number such that $\alpha < \gamma < \beta$.

Conversely, if $F(z)$ is an arbitrary function of class $H_2(\alpha,\beta)$, then, as we proved in §1 of Chapter 11, for any $y \in (a,b)$, where $\alpha < a < b < \beta$, we have the representation (2) of Chapter 11, from which we deduce with the help of the construction used in Chapter 11 that there exists a function $f(t)$ $(-\infty < t < \infty)$ expressing $F(z)$ in the form (3), where $f(t)$ satisfies (1), and $F^+(z)$ and $F^-(z)$ are defined in (2).

We now prove the following useful

PROPOSITION. *Suppose that $F(z) \in H_2(\alpha,\beta)$ and $G(z) \in H_2(\rho,\sigma)$, where $\rho < \sigma < \alpha < \beta$, and that for some $\gamma \in (\alpha,\beta)$ and $\tau \in (\rho,\sigma)$*

$$\frac{1}{2\pi}\int_{i\gamma-\infty}^{i\gamma+\infty} F(z)e^{-itz}\,dz = \frac{1}{2\pi}\int_{i\tau-\infty}^{i\tau+\infty} G(z)e^{-itz}\,dz \stackrel{\text{def}}{=} \omega(t) \qquad (-\infty < t < \infty).$$

*In this case there exists a function $\Omega(z) \in H_2(\rho,\beta)$ that coincides with $F(z)$ in the strip $\alpha < y < \beta$ and with $G(z)$ in the strip $\rho < y < \sigma$, and is thus an analytic continuation of each of these two functions.*

PROOF. Since $F(z) \in H_2(\alpha,\beta)$, what has been proved implies the existence of a function $f(t)$ such that (1) holds and

$$f(t) = \frac{1}{2\pi}\int_{i\gamma-\infty}^{i\gamma+\infty} F(z)e^{-itz}\,dz \qquad (-\infty < t < \infty)$$

for any $\gamma \in (\alpha,\beta)$. Similarly, there exists a function $g(t)$ such that

$$\int_0^{\infty} |g(t)e^{-\rho t}|^2\,dt < \infty, \qquad \int_{-\infty}^{0} |g(t)e^{-\sigma t}|^2\,dt < \infty \tag{1'}$$

and

$$g(t) = \frac{1}{2\pi}\int_{i\tau-\infty}^{i\gamma+\infty} G(z)e^{-itz}\,dz \qquad (-\infty < t < \infty)$$

for any $\tau \in (\rho, \sigma)$. By the condition of the proposition, for almost all $t \in (-\infty, \infty)$ we have $f(t) = g(t) = \omega(t)$. Consequently, $\omega(t)$ satisfies the relations

$$\int_{-\infty}^{0} |\omega(t)e^{-\beta t}|^2\, dt < \infty, \qquad \int_{0}^{\infty} |\omega(t)e^{-\rho t}|^2\, dt < \infty,$$

which follow from (1) and (1′). Therefore,

$$\Omega(z) \overset{\text{def}}{=} \int_{-\infty}^{\infty} \omega(t)e^{itz}\, dt \in H_2(\rho, \beta).$$

Take the numbers $\gamma \in (\alpha, \beta)$ and $\tau \in (\rho, \sigma)$ in the condition of the proposition. Then for all $x \in (-\infty, \infty)$

$$\Omega(x + i\gamma) = \int_{-\infty}^{\infty} \omega(t)e^{it(x+i\gamma)}\, dt = \int_{-\infty}^{\infty} f(t)e^{it(x+i\gamma)}\, dt = F(x + i\gamma),$$
$$\Omega(x + i\tau) = \int_{-\infty}^{\infty} \omega(t)e^{it(x+i\tau)}\, dt = \int_{-\infty}^{\infty} g(t)e^{it(x+i\tau)}\, dt = G(x + i\tau),$$

which yields the proposition, in view of the uniqueness theorem for analytic functions.

**3.** We now again turn to §1 of this chapter in order to consider the case when $\alpha \geq \beta$. In this case the half-planes $y > \alpha$ and $y < \beta$ do not have a common part, and thus the functions (2) cannot be added. However, we can regard the pair $\{F^+(z), F^-(z)\}$ of analytic functions as a certain new object, called the *generalized Fourier transform of the function* $f(t)$. We must prove that the generalized Fourier transform $\{F^+(z), F^-(z)\}$ uniquely determines $f(t)$. To do this we treat the formulas (2) in the way we treated (3) above. Then for any $\alpha' > \alpha$ and $\beta' < \beta$

$$\frac{1}{2\pi} \int_{i\alpha'-\infty}^{i\alpha'+\infty} F^+(z)e^{-itz}\, dz = \begin{cases} f(t) & (t > 0), \\ 0 & (t < 0) \end{cases}$$

and

$$\frac{1}{2\pi} \int_{i\beta'-\infty}^{i\beta'+\infty} F^-(z)e^{-itz}\, dz = \begin{cases} 0 & (t > 0), \\ f(t) & (t < 0). \end{cases}$$

Consequently, the inversion formula has the form

$$f(t) = \frac{1}{2\pi} \int_{i\alpha'-\infty}^{i\alpha'+\infty} F^+(z)e^{-itz}\, dz + \frac{1}{2\pi} \int_{i\beta'-\infty}^{i\beta'+\infty} F^-(z)e^{-itz}\, dz, \qquad (5)$$

where the numbers $\alpha' > \alpha$ and $\beta' < \beta$ are arbitrary.

**4.** The generalized Fourier transform finds an important application in the so-called *Wiener-Hopf technique* for solving integral equations of convolution type on the semi-axis. We confine ourselves to the homogeneous equation, which will be written in the form

$$f(x) = \lambda \int_0^\infty k(x-y)f(y)\,dy \qquad (0 \le x < \infty). \tag{6}$$

The given function (kernel) $k(t)$ is assumed to satisfy

$$|k(t)| \le M \cdot e^{-|t|} \qquad (-\infty < t < \infty). \tag{7}$$

It is also assumed that the unknown function $f(x)$ satisfies

$$e^{-cx}f(x) = O(1) \qquad (x \to \infty) \tag{8}$$

for some fixed $c$ $(0 < c < 1)$. We find the values $\lambda$ for which (6) has nontrivial solutions satisfying (8).

First of all, note that if (6) has been solved, then $f(x)$ can automatically be extended to the semi-axis with the help of this equation. Further, it follows from (6) that for $x < 0$

$$|f(x)| \le |\lambda| M e^{-|x|} \int_0^\infty e^{-y}|f(y)|\,dy,$$

and hence on the whole axis $-\infty < x < \infty$

$$e^{-c|x|}|f(x)| < N.$$

Incidentally, this implies that for any $c' \in (c, 1)$

$$e^{-c'|x|}f(x) \in L^1 \cap L^2(-\infty, \infty).$$

Proceeding to the solution of (6), we introduce the functions

$$F^+(z) = \int_0^\infty f(t)e^{itz}\,dt \in H_2(c', \infty),$$

$$F^-(z) = \int_{-\infty}^0 f(t)e^{itz}\,dt \in H_2(-\infty, -c'),$$

$$K(z) = \int_{-\infty}^\infty k(t)e^{itz}\,dt \in H_2(-c'', c''),$$

where $c < c' < c'' < 1$. Take a $d \in (c', c'')$. By the inversion formula (5),

$$f(t) = \frac{1}{2\pi}\int_{id-\infty}^{id+\infty} F^+(z)e^{-itz}\,dz + \frac{1}{2\pi}\int_{-id-\infty}^{-id+\infty} F^-(z)e^{-itz}\,dz.$$

Since the second integral is equal to zero for $t > 0$, while the first is equal to zero for $t < 0$, we get

$$\begin{aligned}\int_0^\infty k(t-s)f(s)\,ds &= \frac{1}{2\pi}\int_0^\infty k(t-s)\,ds\int_{id-\infty}^{id+\infty} F^+(z)e^{-isz}\,dz\\ &= \frac{1}{2\pi}\int_{-\infty}^\infty k(t-s)\,ds\int_{id-\infty}^{id+\infty} F^+(z)e^{-isz}\,dz\\ &= \frac{1}{2\pi}\int_{id-\infty}^{id+\infty} F^+(z)K(z)e^{-itz}\,dz.\end{aligned}$$

Thus, the left-hand and right-hand sides of (6) have been expressed in terms of the generalized Fourier transform $\{F^+(z), F^-(z)\}$. Substituting the expressions into (6), we arrive at

$$\int_{id-\infty}^{id+\infty} F^+(z)[1-\lambda K(z)]e^{-itz}\,dz + \int_{-id-\infty}^{-id+\infty} F^-(z)e^{-itz}\,dz = 0$$
$$(-\infty < t < \infty). \quad (9)$$

Clearly, $F^+(z)[1-\lambda K(z)] \in H_2(c', c'')$, while $F^-(z) \in H_2(-\infty, -c')$. Applying to (9) the proposition on analytic continuation proved in §2, we find that

$$F^+(z)[1-\lambda K(z)] = -F^-(z) \in H_2(-\infty, c''). \quad (10)$$

This formula must now be used to find $F^+(z)$ in order to obtain $f(t)$ $(0 < t < \infty)$ by the inversion formula. This problem is nontrivial. We postpone its solution until §6, and devote §5 to a simple particular case.

**5.** Consider equation (6) with $k(t) = e^{-|t|}$. Then

$$K(z) = \int_{-\infty}^\infty e^{-|t|}e^{itz}\,dt = \frac{2}{1+z^2} \qquad (-1 < y < 1),$$

and hence

$$1-\lambda K(z) = \frac{1-2\lambda+z^2}{1+z^2} = \frac{(z-z_1)(z-z_2)}{(z-i)(z+i)}, \quad (11)$$

where $z_1 = \sqrt{2\lambda-1}$ and $z_2 = -\sqrt{2\lambda-1}$. Equation (10) takes the form

$$F^+(z)\frac{z^2-(2\lambda-1)}{(z-i)(z+i)} = -F^-(z) \in H_2(-\infty, c''). \quad (10')$$

We confine ourselves to real values of the parameter $\lambda$. Here we distinguish two cases: $(1-c^2)/2 \le \lambda; 2)\lambda < (1-c^2)/2$. In case 1) the points $z_1$ and $z_2$ lie in the strip $-c \le y \le c$, and in case 2) they do not.

In the first case we rewrite (10′) in the form

$$F^+(z)\frac{(z-z_1)(z-z_2)}{z+i} = -F^-(z)(z-i). \quad (\alpha)$$

Since $F^+(z) \in H_2(c', \infty)$, the left-hand side of $(\alpha)$ is regular for $y > c'$. On the other hand, since both sides of $(10')$ and, in particular, $F^-(z)$ belong to $H_2(-\infty, c'')$, the right-hand side of $(\alpha)$ is regular for $y < c''$. Therefore, both sides of $(\alpha)$ are actually regular in the whole plane. Moreover, according to Problem 1 in Chapter 11, the left-hand side of $(\alpha)$ is $o(|z|)$ as $|z| \to \infty$ with $y > c'$, while the right-hand side has this property as $|z| \to \infty$ with $y < c''$. Hence, the left-hand (and right-hand) side of $(\alpha)$ is a constant. Denoting it by $Ci$ and using contour integration, we find that

$$f(t) = \frac{1}{2\pi}\int_{id-\infty}^{id+\infty} F^+(z)e^{-itz}\,dz = \frac{Ci}{2\pi}\int_{id-\infty}^{id+\infty} \frac{(z+i)e^{-itz}\,dz}{(z-z_1)(z-z_2)}$$

$$= C\left[\cos(t\sqrt{2\lambda-1}) + \frac{\sin(t\sqrt{2\lambda-1})}{\sqrt{2\lambda-1}}\right] \qquad (t>0).$$

In particular, for $\lambda = 1/2$

$$f(t) = C(1+t) \qquad (t>0).$$

Passing to the second case, in which $z_1 = i\mu$ and $z_2 = -i\mu$ for some $\mu > c$, we rewrite $(10')$ in the form

$$F^+(z)\frac{z+i\mu}{z+i} = -F^-(z)\frac{z-i}{z-i\mu}. \qquad (\beta)$$

Considering that $c'' < \mu$ and arguing as in the first case, we see that both sides of $(\beta)$ are regular in the whole plane. But the left-hand side of $(\beta)$ tends to zero as $|z| \to \infty$ with $y > c'$, while the right-hand side tends to zero as $|z| \to \infty$ with $y < c''$ (here Problem 1 in Chapter 11 is used again). Consequently, the left-hand (and right-hand) side of $(\beta)$ is identically zero; that is, our equation has only the trivial solution in the second case.

**6.** Analyzing our constructions in the special case, we note that their success is based on the formula (11); that is, on a certain factorization of the function $1-\lambda K(z)$ in (10). Returning to the general case, we present a procedure for getting the factorization of $1-\lambda K(z)$ that is due to Wiener and Hopf and is a basic technique in the solution of the general equation (6).

Observe first of all that in the strip $-c \le y \le c$, where $c < 1$ is given, the function $1-\lambda K(z)$ can have only finitely many zeros for fixed $\lambda$—denote them by $z_1, \dots, z_n$—and it has those zeros in the strip $-c' \le y \le c'$ for some $c' \in (c, 1)$. This follows from the fact that $K(z) \in H_2(-c'', c'')$, and hence $K(x+iy) \to 0$ $(x \to \pm\infty)$ uniformly with respect to $y \in [-c', c']$.

In the strip $-c' \le y \le c'$ the function

$$\frac{1-\lambda K(z)}{(z-z_1)\cdots(z-z_n)}(z+i)^n \qquad (12)$$

does not vanish and tends to 1 uniformly with respect to $y$ as $x \to \pm\infty$. In this strip the limit values of the argument of the quantity in (12) as $x \to +\infty$ and $x \to -\infty$ can be different; however, there exists an integer $m$ such that the limit values of the argument of

$$\psi(z) = \frac{1 - \lambda K(z)}{(z - z_1)\cdots(z - z_n)}(z + i)^n \left(\frac{z - i}{z + i}\right)^m$$

are the same. But in this case the branch of the logarithm can be chosen so that for some $c''$

$$\ln \psi(z) \in H_2(-c'', c'').$$

Then the representation[39]

$$\ln \psi(z) = \frac{1}{2\pi i}\int_{-\infty}^{\infty} \frac{g_1(t)\,dt}{t + ic' - z} + \frac{1}{2\pi i}\int_{-\infty}^{\infty} \frac{g_2(t)\,dt}{t - ic' - z}$$

is valid in the strip $-c' < y < c'$, where $g_1(t), g_2(t) \in L^2(-\infty, \infty)$. Setting

$$\varphi^-(z) = \exp \frac{1}{2\pi i}\int_{-\infty}^{\infty} \frac{g_1(t)\,dt}{t + ic' - z}, \qquad \varphi^+(z) = \exp \frac{1}{2\pi i}\int_{-\infty}^{\infty} \frac{g_2(t)dt}{t - ic' - z},$$

we get the final factorization

$$1 - \lambda K(z) = \frac{(z - z_1)\cdots(z - z_n)}{(z + i)^{n-m}(z - i)^m}\varphi^-(z)\varphi^+(z),$$

which is true in the strip $-c' < y < c'$, where $\varphi^-(z)$ is regular for $y < c'$, does not vanish, and tends to 1 as $|z| \to \infty$, and $\varphi^+(z)$ has the same properties for $y > -c'$. If we then take the equality

$$F^+(z)\varphi^+(z)\frac{(z - z_1)\cdots(z - z_n)}{(z + i)^{n-m}} = -\frac{F^-(z)}{\varphi^-(z)}(z - i)^m,$$

which follows from (10), we see that the everywhere regular function written here is $o(|z|^m)$ as $|z| \to \infty$. Consequently, for $m > 0$ it is a polynomial of degree at most $m - 1$, which we denote by $P_{m-1}(z)$, while for $m \le 0$ it is identically zero. By the inversion formula, for $t > 0$ and for some $d \in (c', 1)$

$$f(t) = \frac{1}{2\pi}\int_{id-\infty}^{id+\infty} \frac{P_{m-1}(z)(z + i)^{n-m}}{\varphi^+(z)(z - z_1)\cdots(z - z_n)}e^{-itz}\,dz \qquad (m > 0);$$

$$f(t) = 0 \qquad (m \le 0).$$

In §5 we have

$$n = 2, \quad m = 1, \quad \varphi^+(z) = 1, \quad \varphi^-(z) = 1$$

---

[39] See formula (2) in Chapter 11.

in case 1), and

$$n = 0, \quad m - 0, \quad \varphi^+(z) = \frac{z + i\mu}{z + i}, \quad \varphi^-(z) = \frac{z - i\mu}{z - i}$$

in case 2).

7. PROBLEMS. 1. Prove that the integral equation

$$f(t) = \lambda \int_0^\infty e^{-|t-s|} f(s)\, ds \qquad (0 \le t < \infty)$$

is equivalent to the differential equation

$$f''(t) - (1 - 2\lambda) f(t) = 0 \qquad (0 \le t < \infty)$$

for the initial condition $f'(0) = f(0)$.

2. Let $\varphi(t) \in L^2(-\infty, \infty)$. Prove that

$$A(z) = \frac{1}{2\delta} \int_{-\infty}^\infty \frac{\varphi(t)}{\cosh(\pi(t - z)/2\delta)} dt \in H_2(-\delta, \delta),$$

$$B(z) = \frac{1}{2\delta} \int_{-\infty}^\infty \frac{\varphi(t)}{\sinh(\pi(t - z)/2\delta)} dt \in H_2(0, 2\delta).$$

*Hint.* The second assertion is a consequence of the first, and the first follows from the representation

$$A(z) = \frac{1}{\sqrt{2\pi}} \int_{-\infty}^\infty \hat{\varphi}(v) \frac{e^{izv}}{\cosh \delta v} dv$$

and the identity (see Problem 2 in Chapter 3)

$$\frac{1}{2\delta} \cdot \frac{1}{\cosh(\pi(t - z)/2\delta)} = \frac{1}{2\pi} \int_{-\infty}^\infty \frac{e^{-i(t-z)v}}{\cosh \delta v} dv \qquad (-\delta < y < \delta).$$

3. Let $f_1(t)$ and $g_0(t)$ be two real functions in $L^2(-\infty, \infty)$. Prove that for the function

$$F(z) = \frac{1}{2\delta} \int_{-\infty}^\infty \frac{f_1(t)}{\cosh(\pi(t - z)/2\delta)} dt + \frac{1}{2\delta} \int_{-\infty}^\infty \frac{g_0(t)}{\sinh(\pi(t - z)/2\delta)} dt \in H_2(0, \delta)$$

and for almost all $x$ $(-\infty < x < \infty)$ the following relations hold:

$$F(x + i0) = \frac{1}{2\delta} \int_{-\infty}^\infty \frac{f_1(t)\, dt}{\cosh(\pi(t - x)/2\delta)} + i g_0(x) + \frac{1}{2\delta} \text{p.v.} \int_{-\infty}^\infty \frac{g_0(t)\, dt}{\cosh(\pi(t - x)/2\delta)},$$

$$F(x+i\delta)=f_1(x)+\frac{i}{2\delta}\,\text{p.v.}\int_{-\infty}^{\infty}\frac{f_1(t)\,dt}{\sinh(\pi(t-x)/2\delta)}+\frac{i}{2\delta}\int_{-\infty}^{\infty}\frac{g_0(t)\,dt}{\sinh(\pi(t-x)/2\delta)},$$

so that

$$\operatorname{Im}F(x+i0)=g_0(x),\qquad \operatorname{Re}F(x+i\delta)=f_1(x).$$

4. (Continuation). Introducing the second pair of functions

$$\operatorname{Re}F(\lambda+i0)=f_0(x),\qquad \operatorname{Im}f(x+i\delta)=g_1(x),$$

we consider on $L^2(-\infty,\infty)$ a linear operator $\mathfrak{G}$ carrying the pair $\{f_1,g_0\}$ into the pair $\{f_0,g_1\}$:

$$\{f_0,g_1\}=\mathfrak{G}\cdot\{f_1,g_0\},$$

if

$$f_0(x)=\frac{1}{2\delta}\int_{-\infty}^{\infty}\frac{f_1(t)\,dt}{\cosh(\tau(t-x)/2\delta)}+\frac{1}{2\delta}\,\text{p.v.}\int_{-\infty}^{\infty}\frac{g_0(t)\,dt}{\sinh(\pi(t-x)/2\delta)},$$

$$g_1(x)=\frac{1}{2\delta}\,\text{p.v.}\int_{-\infty}^{\infty}\frac{f_1(t)\,dt}{\sinh(\pi(t-x)/2\delta)}+\frac{1}{2\delta}\int_{-\infty}^{\infty}\frac{g_0(t)\,dt}{\cosh(\pi(t-x)/2\delta)}.$$

Prove the following facts:

a) The inversion formulas have the form

$$f_1(x)=\frac{1}{2\delta}\int_{-\infty}^{\infty}\frac{f_0(t)\,dt}{\cosh(\pi(t-x)/2\delta)}+\frac{1}{2\delta}\,\text{p.v.}\int_{-\infty}^{\infty}\frac{g_1(t)\,dt}{\sinh(\pi(x-t)/2\delta)},$$

$$g_0(x)=\frac{1}{2\delta}\,\text{p.v.}\int_{-\infty}^{\infty}\frac{f_0(t)\,dt}{\sinh(\pi(x-t)/2\delta)}+\frac{i}{2\delta}\int_{-\infty}^{\infty}\frac{g_1(t)\,dt}{\cosh(\pi(t-x)/2\delta)}$$

i.e.,

$$\{g_0,-f_1\}=\mathfrak{G}\cdot\{g_1,-f_0\}.$$

b) The assumption that the functions are real can be omitted.

c) The operator $\mathfrak{G}$ is unitary; that is,

$$\int_{-\infty}^{\infty}(|f_0(x)|^2+|g_1(x)|^2)\,dx=\int_{-\infty}^{\infty}(|f_1(t)|^2+|g_0(t)|^2)\,dt.$$